AF847579

Springer Series on Atomic, Optical, and Plasma Physics

Volume 118

Editor-in-Chief

Gordon W. F. Drake, Department of Physics, University of Windsor, Windsor, ON, Canada

Series Editors

James Babb, Harvard-Smithsonian Center for Astrophysics, Cambridge, MA, USA

Andre D. Bandrauk, Faculté des Sciences, Université de Sherbrooke, Sherbrooke, QC, Canada

Klaus Bartschat, Department of Physics and Astronomy, Drake University, Des Moines, IA, USA

Charles J. Joachain, Faculty of Science, Université Libre Bruxelles, Bruxelles, Belgium

Michael Keidar, School of Engineering and Applied Science, George Washington University, Washington, DC, USA

Peter Lambropoulos, FORTH, University of Crete, Iraklion, Crete, Greece

Gerd Leuchs, Institut für Theoretische Physik I, Universität Erlangen-Nürnberg, Erlangen, Germany

Alexander Velikovich, Plasma Physics Division, United States Naval Research Laboratory, Washington, DC, USA

The Springer Series on Atomic, Optical, and Plasma Physics covers in a comprehensive manner theory and experiment in the entire field of atoms and molecules and their interaction with electromagnetic radiation. Books in the series provide a rich source of new ideas and techniques with wide applications in fields such as chemistry, materials science, astrophysics, surface science, plasma technology, advanced optics, aeronomy, and engineering. Laser physics is a particular connecting theme that has provided much of the continuing impetus for new developments in the field, such as quantum computation and Bose-Einstein condensation. The purpose of the series is to cover the gap between standard undergraduate textbooks and the research literature with emphasis on the fundamental ideas, methods, techniques, and results in the field.

More information about this series at https://link.springer.com/bookseries/411

Francois Frémont

Classical Treatment of Collisions Between Ions and Atoms or Molecules

Springer

Francois Frémont
University of Caen-Normandie
Caen, France

ISSN 1615-5653 ISSN 2197-6791 (electronic)
Springer Series on Atomic, Optical, and Plasma Physics
ISBN 978-3-030-89427-6 ISBN 978-3-030-89428-3 (eBook)
https://doi.org/10.1007/978-3-030-89428-3

© The Editor(s) (if applicable) and The Author(s), under exclusive license to Springer Nature Switzerland AG 2021
This work is subject to copyright. All rights are solely and exclusively licensed by the Publisher, whether the whole or part of the material is concerned, specifically the rights of translation, reprinting, reuse of illustrations, recitation, broadcasting, reproduction on microfilms or in any other physical way, and transmission or information storage and retrieval, electronic adaptation, computer software, or by similar or dissimilar methodology now known or hereafter developed.
The use of general descriptive names, registered names, trademarks, service marks, etc. in this publication does not imply, even in the absence of a specific statement, that such names are exempt from the relevant protective laws and regulations and therefore free for general use.
The publisher, the authors and the editors are safe to assume that the advice and information in this book are believed to be true and accurate at the date of publication. Neither the publisher nor the authors or the editors give a warranty, expressed or implied, with respect to the material contained herein or for any errors or omissions that may have been made. The publisher remains neutral with regard to jurisdictional claims in published maps and institutional affiliations.

This Springer imprint is published by the registered company Springer Nature Switzerland AG
The registered company address is: Gewerbestrasse 11, 6330 Cham, Switzerland

Preface

Since the beginning of the twentieth century, physics has undergone a major upheaval with the advent of new concepts that were introduced to describe microscopic material properties, on one side, and interactions of massive objects in Nature, on the other side. Very quickly, it has been recognized that classical mechanics, based on the motion principles laid down by Isaac Newton in 1687, was no longer adapted to keep pace with experimental observations. New methods, introducing quantification of various quantities, such as energy or angular momentum, were developed by Heisenberg, Dirac, Schrödinger, to name just a few, and were found to reproduce experiments and to predict new concepts and new particles, such as the Laser effect, the particle spin, or the existence of neutron, positrons, and neutrinos.

Quantum mechanics has been used for many years to describe interactions between particles, and collisions between projectiles and atomic or molecular targets, at low, intermediate, and high projectile energies. For simple systems, such as those involving fully stripped ions and target atoms such as H or He, the total and partial cross sections have been found to nicely reproduce experimental results, regardless of the projectile energy. However, for more complex collision systems, or to further detail a specific collision, strong discrepancies may appear between quantum theories and experiments. Such discrepancies have also been observed in very simple systems, i.e., $H^+ + H$. In conclusion, due to the complexity of quantum mechanics achievement, classical mechanics, at the early stage of research, has been used, first for simple collision systems, then for more complicated systems, involving partly charged ions, multi-electron targets or molecules, as examples. Even if classical mechanics is a priori not adapted for nanoscopic objects, it offers the possibility to visualize the collision itself and to rapidly achieve preliminary results. In addition, classical mechanics is straightforward to implement, thus enabling a comparison with more sophisticated results.

The present book is divided into six sections. Chapter 1 is devoted to the first classical models that were used, before the advent of quantum mechanics, to describe atoms using Bohr and Sommerfeld models and the specific case of He ions (α particles) colliding with gold foils (Rutherford model). In Chap. 2, two classical models are theoretically described. Their advantages and limits will be discussed. In the next chapters, collisions between highly charged ions A^{q+} and atomic or molecular targets are explained in detail. A particular focus will be set on the initial conditions to be met by the target (before the collision) and the challenging treatment of electron correlation, i.e., electron–electron interaction, before, during, and after the collision.

Finally, recent theoretical advances using classical mechanics will be presented. We will show that these calculations cannot be implemented by means of quantum mechanics. It will appear that significant progresses can be made in understanding collision processes and improved by means of classical mechanics.

Caen, France Francois Frémont

Acknowledgments

This book was made possible thanks to the unwavering support of my family, my partner Jocelyne Levillain, and my two daughters, Xavière and Manon, as well as my friends: C. Dufour, Professor of physics at the University of Caen.

I would like to thank my bachelor's and master's professors, who delivered their knowledge and their passion, and helped me finance my thesis and helped me enter the university as Teacher and Researcher: Prof. Cojan, Chantepie, Barat, Vienne.

I would like to thank N. Stolterfoht, my thesis Supervisor and Mentor, who gave me the passion for physics, and R. O. Barrachina, who knew how to instill a powerful and rigorous dynamic on interferences.

I am indebted to my French colleagues (C. Harel, H. Jouin, B. Pons, U. Ancarani, J. Hanssen, P. Moretto-Capelle, A. Bordenave-Montesquieu, and many others), and I thank my colleagues from abroad, whether they are from the USA (J. Tanis, RE Olson), from Hungary (B. Sulik), from Sweden (H. Cederquist), from Belgium (N. Vaeck, H. van der Hart), Argentina (C. Stia, O. Fojon, R. Rivarola, V. Rodriguez, P. Macri), Spain (F. Martin, L. Mendez), and many others.

Many thanks to some colleagues and friends from my laboratory, CIMAP : P. M. Anglade, P. Leprince, D. Hasley, L. Bercker, A. Benayad, L. Adoui, A. Cassimi, F. Stévenot, E. Cagnot, S. Colineaux, S. Duprey, C. Gendreau, H. Gilles, S. Girard, E. Giglio, J. Douaady, J. Rangama, X. Portier, F. Lemarié, and M. Laroche.

I have a thought without any regret for many of my colleagues at the laboratory, as well as at the University of Caen, where I worked for thirty years, who refused to support me in the correction of this book, which was finally carried out by a translator friend of Lille, I. Bertrand, and which prevented my desire to integrate a new team of the laboratory.

In the end, I am infinitely indebted to all my students, to whom we leave a world difficult to integrate and an uncertain future.

Contents

Chapter 1
Classical Mechanics—Goal and First Results

Abstract After the discovery of quantification of matter and energy, quantum physics was introduced very early and developed to describe the structure of atoms and the interactions between particles, and to analyze collision processes during the collision. However, the first models, used in particular cases, were based on classical considerations, and were found to be efficient in understanding, for example, the Rydberg formula from the discretization of atom orbits, or the deviation of α particles in matter.

1.1 The Goal of Classical Mechanics

As mentioned in the preface, classical mechanics, despite its domain of validity in energy, is a useful and simple tool for describing collisions. Even if classical physics avoids fundamental aspects such as quantum numbers, electron spin, discretization of energy, it is well known that quantum mechanics cannot solve all the problems, whatever the high degree of sophistication employed in the available theories. In particular, quantum mechanics is known to be difficult to implement in the case of non-independent electrons. In addition, in view of surprising good results obtained by using classical mechanics, and due to the developments of classical technics to mimic some aspect of quantum mechanics, one expects that interest in using classical mechanics will grow in the near future.

Of course, it has to be recognized that classical mechanics will not replace quantum mechanics, but, at least, it will give preliminary results when quantum mechanics will reach its limitation, essentially due to the complexity of the problems.

© The Author(s), under exclusive license to Springer Nature Switzerland AG 2021

F. Frémont, *Classical Treatment of Collisions Between Ions and Atoms or Molecules*,
Springer Series on Atomic, Optical, and Plasma Physics 118,
https://doi.org/10.1007/978-3-030-89428-3_1

1.2 First Contributions of Classical Mechanics in Atomic Processes

1.2.1 The Rutherford Model

Rutherford experiment and theory follow the Thomson model [1] describing an atom. According to Thomson, an atom consists of a number of corpuscles moving about in a sphere of uniform positive electrification. The corpuscles are constrained to move in one plane, as shown by its calculations, and they will arrange themselves in a series of concentric rings. This arrangement is necessitated by the fact that a large number of corpuscles cannot be in stable equilibrium when arranged in a single ring. In addition, if N electrons occupy the sphere, with a charge equal to $-Ne$ (e is the elementary charge), the charge of the sphere of positive electrification is $+Ne$.

Using this model, it follows that a charged particle that passes through a foil deviates from its incident trajectory with a small scattering angle. Unfortunately, subsequent experiments revealed specific problems using an interpretation with the Thomson model.

Between 1908 and 1913, a series of experiments conducted essentially by Geiger and Marsden [2–5], in which α particles collide with a thin gold foil, disproved Thomson's model. Indeed, in contrast with Thomson predictions, a small part of α particles were found to scatter at large angles, with some going in the direction of the source, i.e. the backward direction. To explain this behavior, the authors encountered an electrostatic force far greater than that allowed for by Thomson's model.

In 1911, E. Rutherford proposed a classical model [6] where the atom consists of mostly empty space, with a positive charge concentrated in its center, and surrounded by a cloud of electrons, kept together by attractive force from the nucleus. Further, the nucleus is assumed to be the seat of the essential part of the mass of the atom, and to have dimensions very small compared with those of the whole atom.

1.2.2 The Bohr Model

In his famous paper of 1913 [7], N. Bohr pointed out difficulties originating from Rutherford theory. Whereas Thomson atom is in a stable equilibrium, it appears, according to the Rutherford model, that certain configuration equilibrium does not exist. Indeed, in Rutherford conception, since an electron has a circular trajectory around the nucleus, this electron loses energy by emitting a radiation, so that it is expected to fall down on the nucleus:

> Let us now, however, take the effect of the energy radiation into account, [...]. In this case, the electron will no longer describe stationary orbits. The kinetic energy of the electron will continuously increase, and the electron will approach the nucleus describing orbits of smaller and smaller dimensions, and with greater and greater frequency; [7].

To circumvent this major difficulty, N. Bohr introduced three postulates:

- An electron that is bound in an atom is able to rotate in certain stable orbits around the nucleus without radiating any energy. These stable orbits are attained at specific discrete distances from the nucleus.
- The stable orbits are reached at distances for which the angular momentum $\ell = m_e vr$ (m_e, v and r are mass, velocity and distance from the nucleus of the electron, respectively) is an integral multiple of the Planck constant, i.e. $\ell = \frac{nh}{2\pi}$.
- Electrons can gain or lose energy by jumping from one allowed orbit to another, absorbing or emitting electromagnetic radiation with a frequency ν given by $E_2 - E_1 = h\nu$, where E_1 and E_2 are the energies of the orbits considered.

Using "ordinary" mechanics, i.e. classical mechanics, but introducing the Planck constant, Bohr was able to determine the famous expression that accounts for the law connecting the lines in the hydrogen spectrum, and to predict series in the extreme ultraviolet and infrared, which cannot be observed directly. In addition, for the hydrogen atom H, the classical formula giving total energy of a bound electron with circular trajectory was derived from the expression of the simple electrostatic force acting on the electron due to the presence of the nucleus. Since the distance r between the nucleus and the electron, the electron velocity v and its energy E are unknown, three equations are required. For an atom with atomic number Z, the total energy is expressed, in the atomic unit system[1]:

$$E = E_c + E_p = \frac{1}{2}v^2 - \frac{Z}{r} \tag{1.1}$$

The force acting on the electron due to the presence of the electron is for acircular orbit.

$$F = \frac{1}{r^2} = \frac{v^2}{r} \tag{1.2}$$

Finally, the angular momentum is quantized, leading to

$$\ell = vr = n \tag{1.3}$$

From the above expressions, it was easy to demonstrate that

$$E = -\frac{Z^2}{n} \tag{1.4}$$

For atomic hydrogen H (Z = 1), the exact value of fundamental level total energy of −13.6 eV could be deduced.

[1] In the atomic unit system, $e = 1$; $m_e = 1$; $\hbar = 1$; $1/4\pi\varepsilon_0 = 1$.

1.2.3 Sommerfeld Model

Sommerfeld's answer to Bohr's atomic model was the earliest reaction from outside Rutherford's model. Sommerfeld pointed out some difficulties related to Bohr's model. The first one is the expression of a classical quantity (Rydberg constant) in terms of the Planck's constant h, that was precisely introduced to describe non-classical effects, namely radiations of black bodies. In a second point, Sommerfeld and his contemporaries were troubled by the fact that the Bohr quantum atom appeared to require circular orbits only and, in particular, by the consequence that, for any fixed electron energy, there is only one possible value of the corresponding angular momentum ([8] and references therein). The third difficulty was the interpretation of the Zeeman effect and the Stark effect using Bohr's model. We recall that these effects describe the splitting of the spectral lines of hydrogen in a magnetic field and an electric field, respectively.

By analogy with Kepler equations, in which elliptic orbits appear, Sommerfeld, considered electron orbits as elliptic with angular momenta ℓ depending on the principal quantum number n. Sommerfeld introduced an integer m, verifying $-n < m < +n$ for a fixed value of n. As an example, some ellipses are represented in Fig. 1.1 for $n = 4$. The different values for m give rise to specific eccentricities of the ellipses.

With the quantization of elliptical orbits and the introduction of relativistic mechanics, Sommerfeld could derive a more complex expression of the bound electron energy than that obtained by Bohr

$$E = E_n\left[1 + \frac{\alpha^2 Z^2}{n^2}\left(\frac{n}{\ell + 1} - \frac{3}{4}\right)\right] \tag{1.5}$$

where α is the fine-structure constant and ℓ a new quantum number, characterized by $\ell = 0, 1, \ldots, n - 1$.

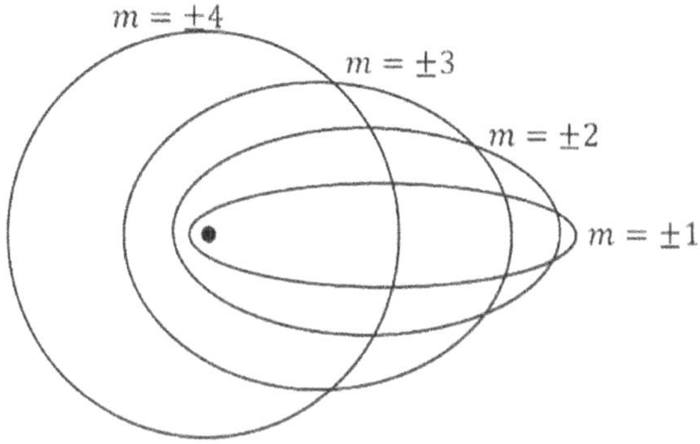

Fig. 1.1 Sommerfeld elliptical orbits for $n = 4$

Unfortunately, despite all these efforts, some other effects, such as abnormal Zeeman effect, describing the splitting of spectral lines of hydrogen in a weak magnetic field, could not be explained by the Sommerfeld model.

As a first conclusion, classical mechanics was able to give good results at the beginning of the discovery of the atomic properties, but only with the introduction of discrete numbers, revealing the necessity of applying a kind of quantization.

References

1. Thomson, J.J.: On the structure of the atom: an investigation of the stability and periods of oscillation of a number of corpuscles arranged at equal intervals around the circumference of a circle; with application of the results to the theory of atomic structure. Phys. Mag. Ser. **6**, 237 (1904)
2. Geiger, H.: On the scattering of α particles by matter. Proc. Roy. Soc. London **81**, 546 (1908)
3. Geiger, H., Marsden, E.: On a diffuse reflection of the α particles. Proc. Roy. Soc. London **82**, 557 (1909)
4. Geiger, H.: The scattering of α particles by matter. Proc. Roy. Soc. London **83,** 565 (1910)
5. Geiger, H., Marsden, E.: The laws of deflection of α particles through large angles. Philos. Mag. **25**, 148 (1913)
6. Rutherford, E.: The scattering of α and β particles by matter and the structure of the atom. Philos. Mag **21**, 669 (1911)
7. Bohr, N.: On the constitution of atoms and molecules. Philos. Mag. **26**, 1 (1913)
8. Deeney, T., O'Sullivan, C.: Sommerfeld's elliptical atomic orbits revisited—A useful preliminary to the study of quantum mechanics. Am. J. Phys. **82**, 9 (2014)

Chapter 2
Classical Mechanics' Approaches in Atomic Collisions

Abstract Before entering into the details of the Classical Trajectory Monte Carlo (CTMC) model, a few older classical models are presented. The Binary Encounter model is first introduced. Then the Over-Barrier model (OBM) is treated into details. Its advantages are discussed, and we will show that electron correlation cannot be understood within this model. We will present some extensions of this model. Then, CTMC model is presented into details. From equations of motions, and model representations of potential energies, expressions for total, partial and differential cross sections will be established.

2.1 Binary Encounter Models

Since the development of quantum mechanics theories, classical methods were largely neglected until 1959. As mentioned by M. Gryzinski at this date [1], "the difficulty of explaining on the basis of classical mechanics some experimental facts in atomic collisions have been viewed as a proof of the non-validity of classical mechanics ..." for atomic size systems. Consequently, for systems involving more than two particles, inelastic processes, such as ionization, excitation or electron capture have not been investigated to give a classical interpretation of these processes.

Gryzinski took advantage of good agreement between previous classical calculations for stopping power in matter with experiment to suggest that other processes should be treated in this way. However, since the problem to solve is not a two-body problem, but at least a three-body problem, involving a projectile nucleus and an atom, constituted by a nucleus and one or more than one electron, the idea has been to transform it to a two-body process. Thus, the target nucleus is supposed not to participate to the collision, so that the projectile only collides with the target electron (Fig. 2.1). This process is called Binary Encounter (BE) process.

Consequently, the basic idea of the BE collision theory is that ionization, excitation, and electron capture are approximately described as a collision between two particles. Also, in Gryzinski's model, the treatment of the two interacting particles is unsymmetrical. For instance, in a collision between an electron and an atomic

© The Author(s), under exclusive license to Springer Nature Switzerland AG 2021

F. Frémont, *Classical Treatment of Collisions Between Ions and Atoms or Molecules*, Springer Series on Atomic, Optical, and Plasma Physics 118,
https://doi.org/10.1007/978-3-030-89428-3_2

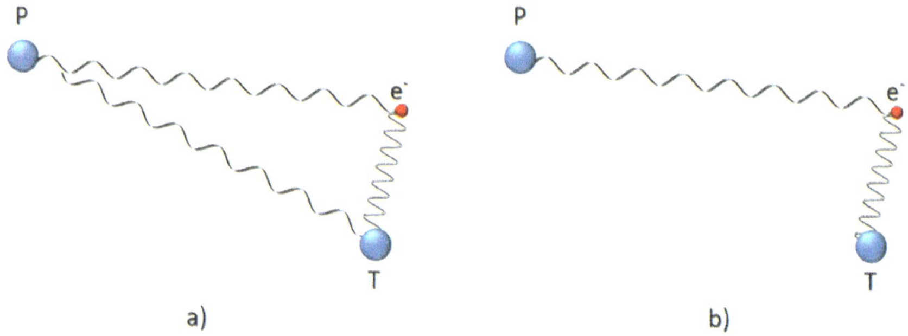

Fig. 2.1 Binary Encounter schematic view. In **a**, all the interactions are taken into account. In **b**, the projectile only interacts with the active electron

hydrogen target, if the incident particle and the atomic electron have the same distance r from the target nucleus, it is assumed that, because of the interaction with the nucleus, the atomic electron has potential energy $-1/r$, whereas the incident particle has potential energy 0.

2.1.1 *Brief Description of the Model and Results*

Consider the collision between an incident particle of mass m_1, charge q_1 and velocity $\overrightarrow{v_1}$ and a target electron of mass m_2, charge q_2 and velocity $\overrightarrow{v_2}$ (Fig. 2.2). The vectors $\overrightarrow{v_1}$ and $\overrightarrow{v_2}$ define the fundamental plane (P_f), and the orbit of the target electron defines the orbital plane (P_o). We note $\Delta E = E_1' - E_1$, where E_1 and E_1' are the initial and final projectile energies, respectively, and θ the change in the direction of the projectile velocity, also called scattering angle. The cross section $\sigma(\Delta E)$ and $\sigma(\theta)$ for producing a change in energy and in angle, respectively, are given by

$$\sigma(\Delta E) = \frac{1}{4\pi} \int f(\theta, \alpha, b)\delta(h_{\Delta E})d\theta d\alpha d\varphi b db \tag{2.1}$$

$$\sigma(\theta) = \frac{1}{4\pi} \int f(\theta, \alpha, b)\delta(h_{\theta})d\theta d\alpha d\varphi b db \tag{2.2}$$

In the above expressions, α is the angle between the orbital plane and the fundamental plane, b is the impact parameter, $f(\theta, \alpha, b)$ is the relative angle distribution between $\overrightarrow{v_1}$ and $\overrightarrow{v_2}$. The quantities $h_{\Delta E}$ and h_θ refer to relations fulfilled by ΔE and θ, respectively (for more details, see [1]). In the specific case where the collection of projectiles have an isotropic distribution of velocities, $f(\theta, \alpha, b)$ writes

$$f(\theta, \alpha, b) f(\theta) = \frac{V sin\theta}{v_1} \tag{2.3}$$

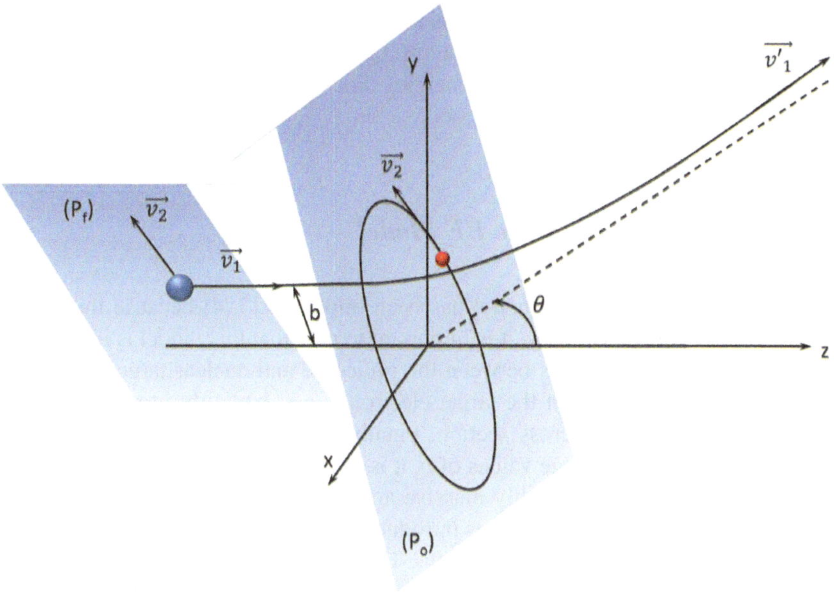

Fig. 2.2 Schematic view of the collision between a projectile with incident and final velocities $\overrightarrow{v_1}$ and $\overrightarrow{v_1'}$, respectively and a target electron of velocity $\overrightarrow{v_2}$. (P_f) and (P_o) are the fundamental and orbital plane, respectively, b is the impact parameter, θ is the azimuthal angle

where $V = \left(v_1^2 + v_2^2 - 2v_1 v_2 cos\theta\right)^{1/2}$ is the relative velocity.

Consequently, after integration of expressions (2.1) and (2.2).

The present theory was applied by Gryzinski for electron projectiles impacting on various targets. As an example, Fig. 2.3 shows calculated total ionization cross sections σ_I (full curves) following $e^- + He$ and $e^- + H_2$ collisions as a function

Fig. 2.3 Total ionization cross sections following e^- + He and e^- + H_2 collisions as a function of the projectile energy. Experiment, full squares, and circles [2]; calculations by Gryzinski, full curves [1]

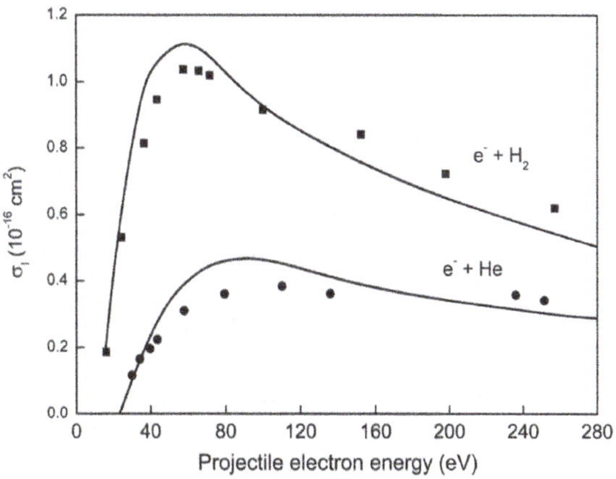

of incident projectile energy. The agreement with experiment [2] is rather good, especially in the shape of the cross sections. Quantitatively, disparities are observed at maximum and at large projectile energies. For other multielectronic targets, the agreement is also found to be quite good.

2.1.2 Improvements of the BE Model

For many years, Gryzinski's model has been improved [3, 4] because this model has deep disadvantages. Firstly, it is important to note that this model is unsymmetrical since no potential exists between the projectile and nuclear target. Secondly, the ionization potential I_p of the target electron is not explicitly introduced in the expressions of ionization cross section. Finally, BE model is only valid at small impact parameters b. At large values of b, it is clear that the projectile may interact with the target nucleus for highly massive and charged ions and electrons at high projectile energies. A new model was introduced by Burgess [3], with a symmetrical treatment of the two interacting electrons. He assumed that the electron projectile gains a kinetic energy I_p and simultaneously loses the same amount of potential energy before it interacts with the target electron which is assumed to be bounded to the nucleus with ionization potential I_p. If the Burgess' model gives better agreement than Gryzinski's model at small incident projectile energies, it does not improve the cross sections at high energies.

Ionized particle-atom collisions can be separated into two types: soft collisions (large values of b) and hard od binary collisions (small values of b). Gryzinski's model only takes into account hard collisions. It has been shown by Bethe [5] that soft collisions give rise to low-energy electrons, whose production is dominant, and that their ejection is due to dipole transitions similar to those produced by photons. Due to the mixing of soft and hard collisions, combination of BE model and dipole models is not obvious to treat analytically. Some approaches to represent σ_I in analytic forms unfortunately require parameters that must be used to fit some theoretical or experimental ionization cross sections. In addition, in the particular case of electron-atom collisions, since projectile and target electrons are undistinguishable, exchange effect has to be included.

A complete expression of σ_I has been developed in 1994, using a binary-encounter-dipole (BED) model, for electron impact ionization. To describe dipole interaction, the asymptotic expression of σ_I in the Bethe theory has been taken, in which the differential oscillator strength is introduced. Since this quantity strongly depends onto the collision system, and even if the BED model is simpler to use than most of ab initio theories, it is often difficult to get differential oscillator strengths. Thus, for such cases, a simplified version of the BED model has been given, in which a simple function is used for differential oscillator strengths [4].

Figure 2.4 and show σ_I for $e^- + H$ collisions as a function of projectile energy. Total cross sections derived from BEB (full curves), and BED (dashed curves) models are compared with Gryzinski's model (short dashed curve) and distorted wave Born

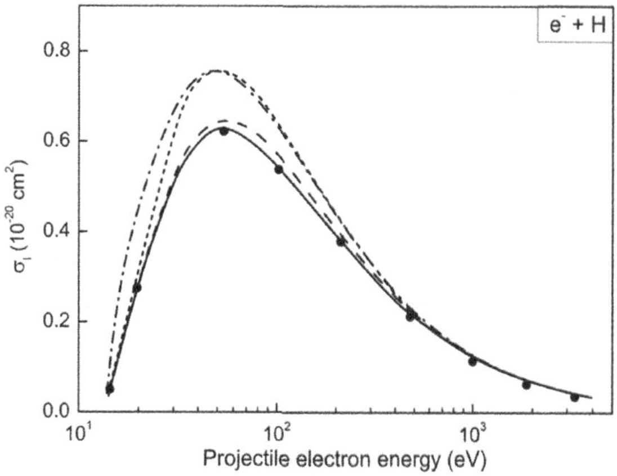

Fig. 2.4 Total ionization cross section σ_I for $e^- + H$ collisions as a function of the projectile energy. Experiment, full circles [7]. Calculations: short dashed curve, Gryzinski's model [1]; dashed curve, Born model [6]; dashed curve, BED model [4]; full curve, BEB model [4]

cross sections (dotted-dashed curve) [6]. The present BEB and BED models are in very good agreement with experiment [7], whereas the two other models are far to fit experiment, especially at energies smaller than 300 eV.

In the case of $e^- + H_2O$ collisions (Fig. 2.5), where the target is molecular and has many electrons, it is seen that the BEB model is very powerful to reproduce experiment. Note that the BED model was not used because oscillator strengths are not available.

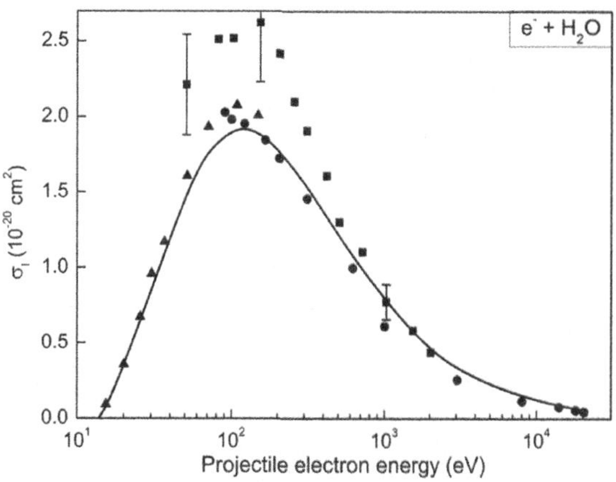

Fig. 2.5 Total ionization cross section σ_I for $e^- + H_2O$ collisions as a function of the projectile energy. Experiment, full triangles [8]; full squares [9]; full circles [10]. Theory, full curve, BEB model [4]

2.2 Classical Over-Barrier Model

2.2.1 Introduction

While classical BE model is mainly devoted to ionization following electron and atom or molecules, a classical over-barrier (COB) model was introduced in 1980 [11] to examine experimental single electron capture (or charge transfer) cross sections in collisions of partially stripper ions with H at low collision energies. This model was created because good agreement was shown between experiments and calculations at impact energies larger than 10 keV/amu whereas, at smaller energies, differences appear, especially for projectiles of ionic charges 2 and 3 [12].

2.2.2 Principle of the Model for Single Electron Capture

In the present model, electron capture corresponds to the transfer of an electron having a sufficient energy to overcome the potential barrier the projectile ion A^{q+} and the target nucleus H^+ (Fig. 2.6).

The potential at internuclear distance R (distance between both nuclei) is written as

$$V(x) = -\frac{q}{x} - \frac{1}{R - x} \tag{2.4}$$

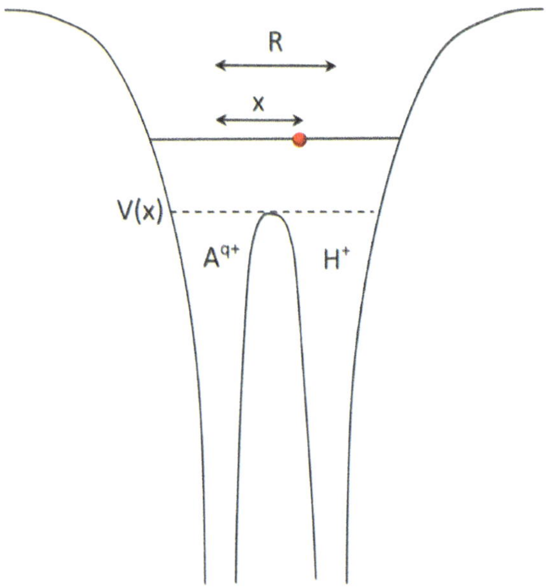

Fig. 2.6 Potential energy of an electron located at distances x of a projectile $A^{q+}$and $R - x$ of the nuclear target. The quantity $V(x)$ is the height of the classical barrier. The electron is not attached to the projectile or the target but moves in the molecule formed by both nuclei

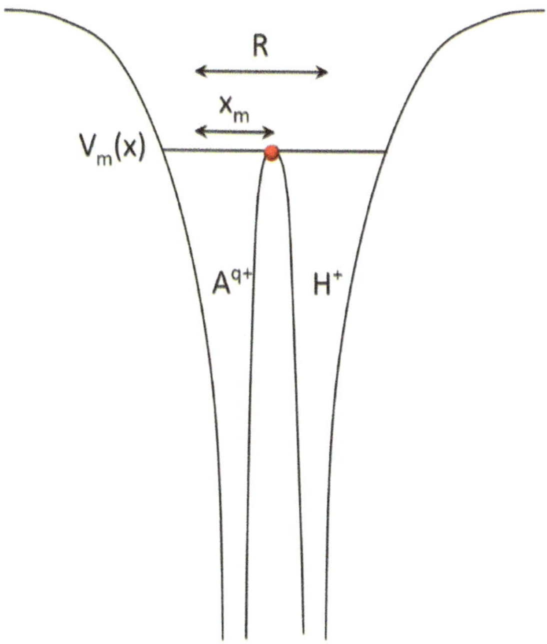

Fig. 2.7 Potential energy of an electron located at distances x_m of a projectile $A^{q+}$and $R - x_m$ of the nuclear target. The quantity V_m is the maximum of the height of the barrier, for which the target electron can be molecularized

where x is the distance between the projectile and the electron that can be captured. While the electron is not captured, it belongs to the molecule formed by the projectile and the target nuclei. The electron is called molecularized.

The maximum of $V(x)$ is obtained by derivation of the above expression, and is given by (Fig. 2.7)

$$V_m = -\frac{(\sqrt{q}+1)^2}{R} \tag{2.5}$$

The electron can be captured by the projectile if the two following conditions are fulfilled. First, the electron energies on the projectile and on the target are equal (resonance condition). Second, the electron energy has to be larger than V_m. This leads to

$$-\frac{1}{2} - \frac{q}{R} = -\frac{q^2}{2n^2} - \frac{1}{R} \tag{2.6}$$

$$-\frac{1}{2} - \frac{q}{R} > V_m \tag{2.7}$$

From relations (2.6) and (2.7), the principal quantum number n on the projectile satisfies

$$n \leq n_p = \left(\frac{2\sqrt{q}+1}{q+2\sqrt{q}}\right)^{1/2} \tag{2.8}$$

and the corresponding internuclear distance at which electron capture occurs is

$$R_n = n^2 \frac{2(q-1)}{q^2 - n^2} \tag{2.9}$$

In this model, n_p is the most probably occupied state of the projectile. It is seen that, since R_n is a positive quantity, the maximum value of n cannot be larger than q.

The single capture cross section σ_{SC} is then determined in terms of R_n, by assuming the capture probability to be 1/2

$$\sigma_{SC} = \sum_{n=1}^{n_p} \frac{1}{2} \pi R_n^2 \tag{2.10}$$

The COB model has many advantages. First, it is a very simple model that can be applied to many collision systems. Second, it predicts quite well the final quantities n, R_n and σ_{SC}. As examples Figs. 2.8 and 2.9 show the principal quantum number and the total cross sections as a function of q. The results agree well with quantum mechanics results [4]. The agreement is particularly visible for σ_{SC}, since both predict strong oscillatory structures. These oscillations are also observed experimentally (see for example [13]), and calculations are seen to coincide very well with experiment.

Despite these remarkable results, the COB model, due to its simplicity, many disadvantages.

- It is a one-dimension model, since it depends only on the variable x.
- It is a one-electron model. Thus, multi-electron capture cannot be studied.

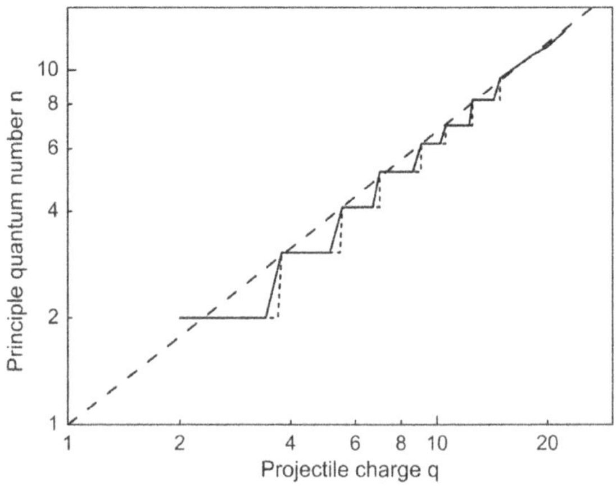

Fig. 2.8 Most probable principal quantum number n of the final states for the charge transfer in A^{q+} + H(ls) collisions. The solid line denotes the values obtained by interpolations from plots of the UDWA partial cross sections at the ion impact energies of 0.1 keV/amu [11]. The dotted line shows the results obtained by the classical treatment

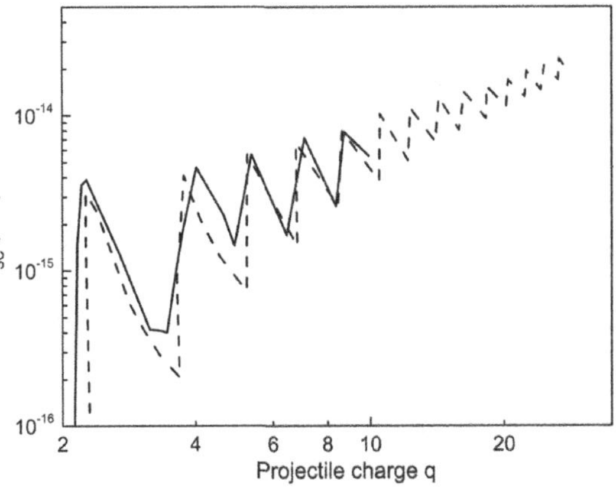

Fig. 2.9 Single electron capture cross sections as a function of the projectile charge q, following A^{q+} + H(ls) collisions. The solid line denotes the UDWA results [11], at the ion impact energy of 0.1 keV/amu, and the dotted line the classical results

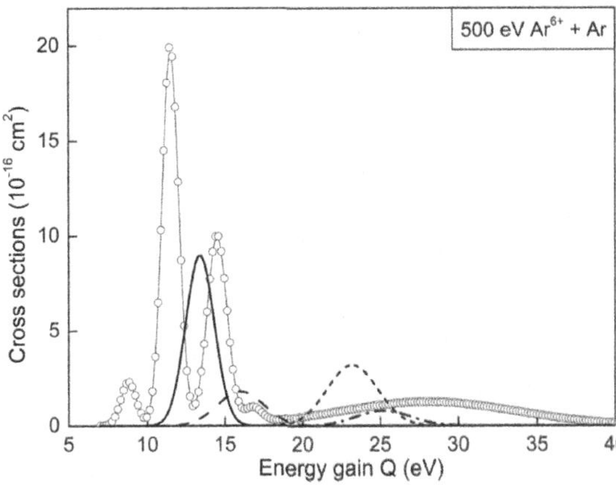

Fig. 2.10 Energy gain distributions following electron capture in 500 eV Ar^{6+} + Ar. Empty circles, experiment [16]; present calculations: solid curve, string (1,0); dashed curve, string (0,1); double capture, short dashed curve, string (1,1); transfer ionization, dotted dashed curve, string (1,0,1)

- The n distribution is available, but is not realistic, since the maximum of n is also the most probable value, which is not verified experimentally.
- There is no dependence of σ_{SC} with projectile velocity, so that this model is only valid at low energy, where σ_{SC} is nearly constant.
- Only total cross sections can be evaluated. Differential and partial cross sections are excluded from the calculation.

2.2.3 *Multi-Electron Capture*

Suppose a projectile A^{q+} of charge q colliding with a multi-electronic target B. The collision writes

$$A^{q+} + B \rightarrow A^{(q-r)+} + B^{(r+n)+} + ne^- \tag{2.11}$$

After the collision, r electrons are captured by the projectile and stay on it, while n electrons are ejected by autoionization. A first extension of SC to MC has been formulated by Barany [14]. This extended model allowed absolute cross sections σ_{SC} to be calculated. However, it did not take into account for autoionization of projectile or target.

A more detailed classical model was formulated in 1986 [15], in order to allow more predictions to be made than those using Barany's model. During the collision, two distinguishable parts are considered: the way in and the way out. On the way in, the internuclear distance R is sufficiently small so that all target electrons have been molecularized. On the way out, the height of the barrier increases, so that r electrons are captured while the others remain on the target. The model defines strings (j) to each capture process. For example, $(j) \equiv (1, 0, 1)$ means that the first and the third electrons are captured, whereas the second electron remains on the target. Consequently, to treat double capture with a target that has three electrons, three strings are possible.

For a given number of capture electrons $(q - r)$, capture cross section is summed over the different strings

$$\sigma_{q-r} = \sum_{(j)} \sigma_{q-r}^{(j)} \tag{2.12}$$

with

$$\sigma_{q-r}^{(j)} = \sum_{t=1}^{t_{max}} \pi\left[\left(R_t^i\right)^2 - \left(R_{t+1}^i\right)^2\right] P_t^{(j)} \tag{2.13}$$

In expression (2.13) $\pi\left[\left(R_t^i\right)^2 - \left(R_{t+1}^i\right)^2\right]$ is the area of the ring-shaped surface of radii between R_t^i and R_{t+1}^i, where i is the index of electron (i) in the string (j). The quantity $P_t^{(j)}$ depends onto the probability $W_t^{(j)}$ of capture. Assuming for $W_t^{(j)}(E)$ or $W_t^{(j)}(Q)$ (E and Q are the binding energies and energy gain, respectively) Gaussian distributions around the most probable values of E and Q, absolute total cross sections and partial cross sections for energy gain could be formulated and calculated.

As an example, for Ar^{6+} + Ar collisions at a collision energy of 500 eV, experimental energy gain distributions [16] are represented in absolute scale (empty circles) and compared with present calculations, that are shown for the main contributions (full curves). The strings (10) and (01) coincide rather well with experiment, and strings (11) and (101) correspond, at energy gain values larger than 20 eV, to Transfer Ionization (TI) process. However, strong discrepancies were observed in terms of cross sections (Fig. 2.10).

Many extensions were performed to treat or to improve particular collision systems. Appendix A presents two examples of such extensions. However, despite all these efforts, COB model is unable to correctly predict final n distributions in collisions of ions with He target at very low impact velocities. Indeed, in this case, electron correlation plays a fundamental role and, whereas COB model is independent onto the projectile velocity, experiments clearly show shifts in the final state distributions. Moreover, total cross sections strongly differ from experiment, even for ions colliding on atomic hydrogen.

2.3 Classical Trajectory Monte Carlo Method

2.3.1 *Origin of the Method and Preliminary Applications*

The principle of the Monte Carlo simulation is to build and use a model that theoretically describes the evolution of a system, estimating one or several quantities from random initial numbers. In this way, this method simulates an experiment. The name of Monte Carlo was introduced by Stanislaw Ulam and John von Neumann, in reference to the famous games of chance in casinos.

The Monte Carlo method has a double advantage: it is very simple to use, and it can be applied to a vast variety of situations. This method was introduced in chemistry in the middle of the 1930's to treat reactions involving molecules and atoms [17]. In fact, the procedure is semi classical, in the sense that potential surfaces are calculated from first-order perturbation theory derived from quantum mechanics formalism. From this potential energy surfaces, trajectories could be calculated point by point from the classical equations of motion.

The method was performed in this field in order to increase the understanding of the dynamics of chemical reactions, through calculations of trajectories involving three or more atoms, and the evaluation of cross sections with more refined precision. It was for example applied with success for $H + H_2$ collisions [18, 19].

The classical Monte Carlo trajectory (CTMC) method was created in 1966 [20] to treat dynamics of ion-atom collisions at relatively high energies and to determine cross sections for single charge exchange and single ionization. The quantum system is represented by a macroscopic classical model. The method is called exact by the authors "*in the sense that no significant dynamical approximations are made*" [20]. However, one has to be careful since Coulombic potentials that are used are approximations, especially at very small distances between the collision partners.

2.3.2 *Brief Description of CTMC Method*

Suppose an ion A^{q+} of mass M_A and charge q colliding on a target B of mass M_B, constituted by N possible active electrons. The CTMC method can be divided in several independent steps (Fig. 2.11).

- It is first necessary to build the target B and the projectile A. This requires mainly the choice of adequate potential energies (cf. next chapters) for the electrons.

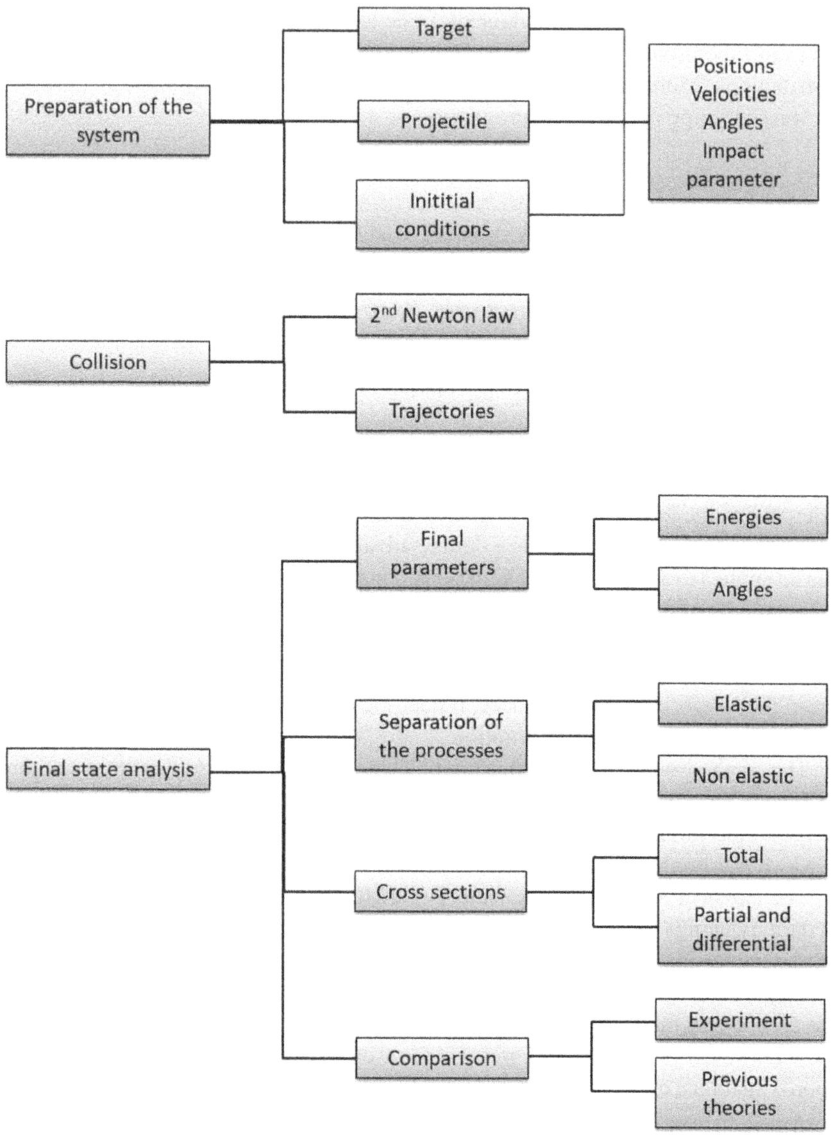

Fig. 2.11 Summary of the different steps in CTMC calculations. The first step consists of the preparation of the system. The second step is the collision itself, and the third step is the analysis of the different quantities that are relevant after the collision

- The collision depends on the initial conditions, for the projectile and the target. One has to properly define the orientation of the target, the projectile velocity, the impact parameter, and the position in space of the projectile.
- Thus the fundamental principle of dynamics is applied, in order to follow the evolution of the collision partners, characterized by trajectories.
- At the end of the collision, all the final parameters (energies, momenta, angles) can be evaluated, with their uncertainties.
- Elastic and inelastic processes have to be separated, according to the finale energies of the electrons involved during the collision.
- Total, partial and differential cross sections are deduced from the number of trajectories for each process.
- Finally, a comparison with other previous theories and experiments has to be made, in order to validate the method.

2.3.3 Equations of Motion and Preparation of the System

In this section, the goal is not to use the chronology of the different steps mentioned above, but rather to give the most general features, then to enter into the details.

2.3.3.1 Equations of Motion

We suppose that all the particles (nuclei and electrons) are punctual, and that each particle interacts with the other one via Coulombic interaction. The force that acts on particle j due to the presence of particle i is

$$\overrightarrow{F}_j(i) = \frac{1}{4\pi\epsilon_o}\frac{q_i q_j}{r_{ij}^2}\overrightarrow{u_{r_i}} \tag{2.14}$$

In the atomic unit system, in which $\frac{1}{4\pi\epsilon_o}$, the elementary charge e, the electron mass and the quantity $\hbar$ are equal to unity, the previous relation writes

$$\overrightarrow{F}_j(i) = \frac{q_i q_j}{r_{ij}^2}\overrightarrow{u_{r_i}} \tag{2.15}$$

In the above expression, q_i is the charge or particle i, r_{ij} is the distance between particles i and j, and $\overrightarrow{u_{r_i}}$ is a unitary vector.

Each particle is, in principle, subjected to the forces due to the presence of the other particles. Applying the second Newton law, we obtain

$$\sum_{j\neq i}\overrightarrow{F}_j(i) = q_i\sum_{j\neq i}\frac{q_j}{r_{ij}^2}\overrightarrow{u_{r_i}} = M_i\overrightarrow{a_i} \tag{2.16}$$

where M_i and $\overrightarrow{a_i}$ are the mass and the acceleration of the particle i, respectively.

However, (2.16) has the disadvantage that it is a second order equation to compute. Thus, the easiest way to treat the problem is to solve Hamilton's equations.

The expression of the total classical Hamiltonian is

$$H = \sum_{i=1}^{N+2} \frac{1}{2} M_i v_i^2 + \sum_{j \neq i} V_{ij} \tag{2.17}$$

where the first term describes the sum of the kinetic energies of each particle, and the second term is the sum of the interaction potential energies between two particles. In the first approximation, Coulombic potentials are sufficient; however, we will see that specific interactions require the necessity to develop particular expressions of potentials.

The equations of motion are

$$\begin{aligned} \frac{\partial r_{i\alpha}}{\partial t} &= \frac{\partial H}{\partial p_{i\alpha}} \\ \frac{\partial p_{i\alpha}}{\partial t} &= -\frac{\partial H}{\partial r_{i\alpha}} \end{aligned} \tag{2.18}$$

In expressions (2.18), the quantities i and α refer to the $N + 2$ particles and to the coordinates x, y, and z, respectively. The quantities $r_{i\alpha}$ and $p_{i\alpha}$ are the coordinates of position and momentum of each particle, respectively.

2.3.3.2 Initial Conditions

To simplify the following discussion, we consider the simplest collision system $A^{Z+} + H$, involving a fully stripped projectile ion and a unique active electron. Consequently, we have to follow three bodies, consisting in two nuclei and an electron. Before the collision, the projectile is sufficiently far from the target, so that its interaction with H can be considered as negligible.

With the condition $r_{eT} = 1$a.u. (r_{eT} is the distance between the active electron and the target nucleus), the initial distance z_o between both nuclei is of the order of 100 a.u.

The coordinates of the projectile and the square of the impact parameter b are randomly chosen. The initial coordinates of the projectile are

$$\begin{cases} x_{p_o} = b\cos\theta_{p_o} \\ y_{p_o} = b\sin\theta_{p_o} \\ z_{p_o} = z_o \end{cases} \tag{2.19}$$

The azimuthal angle θ_{p_o} is randomly chosen between 0 and 2π.

The target nucleus is located at the origin of the coordinate system, and the position of the active electron is randomly selected, with the unique condition that the energy of the electron is—0.5 a.u. In spherical coordinates, the position is

$$\begin{cases} x_{e_o} = r_{e_o}\sin\theta_{e_o}\cos\varphi_{e_o} \\ y_{e_o} = r_{e_o}\sin\theta_{e_o}\sin\varphi_{e_o} \\ z_{e_o} = r_{e_o}\cos\theta_{e_o} \end{cases} \tag{2.20}$$

In order to obtain a uniform distribution, the angle φ_{e_o} and the quantity $\cos\theta_{e_o}$ are randomly chosen between 0 and 2π and -1 and 1, respectively.

With the knowledge that the electron trajectory around the target nucleus is elliptic, the electron velocity is not necessarily orthogonal to the position vector. Consequently, an additional angle η is introduced, varying from $-\pi$ to π. Initial coordinates of electron momentum are

$$\begin{cases} p_{x_{eo}} = -p_{e_o}\left(\sin\varphi_{e_o}\cos\eta + \cos\theta_{e_o}\cos\varphi_{e_o}\sin\eta\right) \\ p_{y_{eo}} = p_{e_o}\left(\cos\varphi_{e_o}\cos\eta - \cos\theta_{e_o}\sin\varphi_{e_o}\sin\eta\right) \\ p_{z_{eo}} = p_{e_o}\sin\theta_{e_o}\sin\eta \end{cases} \tag{2.21}$$

2.3.4 *Spatial and Momentum Initial Distributions*

2.3.4.1 Quantum Distributions

In the frame of quantum mechanics, the spatial distribution is given by the quantity $r_e^2|\psi(r_e)|^2$, where $\psi(r_e)$ is the wave function associated with the electron in its initial (fundamental or excited) state

$$r_e^2|\psi(r_e)|^2 = \alpha r_e^2 e^{-2\frac{r_e}{r_{eo}}} \tag{2.22}$$

where α is a normalization constant.

In the same way, the momentum distribution is

$$p_e^2|\psi(p_e)|^2 = \beta p_e^2 \frac{1}{\left[p_e^2 - p_{e_o}^2\right]^2} \tag{2.23}$$

where β is also a normalization constant.

The radial and momentum densities, describing the electron in its fundamental state on H, with a unique energy of –0.5 a.u., are represented in Figs. 2.12 and 2.13, respectively.

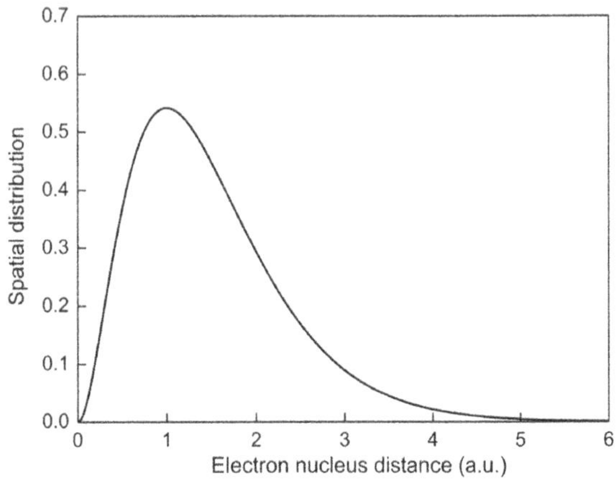

Fig. 2.12 Quantum spatial distribution of hydrogen electron in its fundamental state

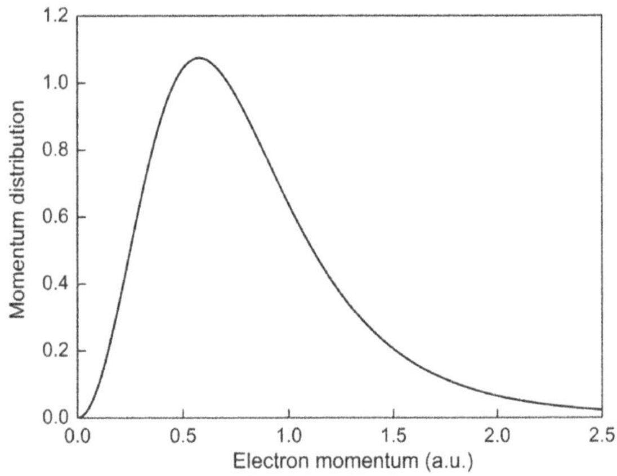

Fig. 2.13 Quantum momentum distribution of hydrogen electron in its fundamental state

2.3.4.2 Classical Distributions

The most useful classical distributions were those described by Abrines and Percival [20]. In these calculations, it is assumed that the initial coordinates and momenta of the electron bound to the proton are distributed in the phase space of the electron according to the microcanonical distribution. For the 1s electron in the Coulomb field of the proton, it yields to the correct quantum mechanical density, but not the density in coordinate space.

In phase space, the distribution is, for an energy E_o

$$\rho_{E_o}\left(\vec{r_e}, \vec{p_e}\right) = \delta\left(E_o - \frac{p_e^2}{2} - \frac{1}{r_e}\right) \tag{2.24}$$

where δ is the Dirac distribution.

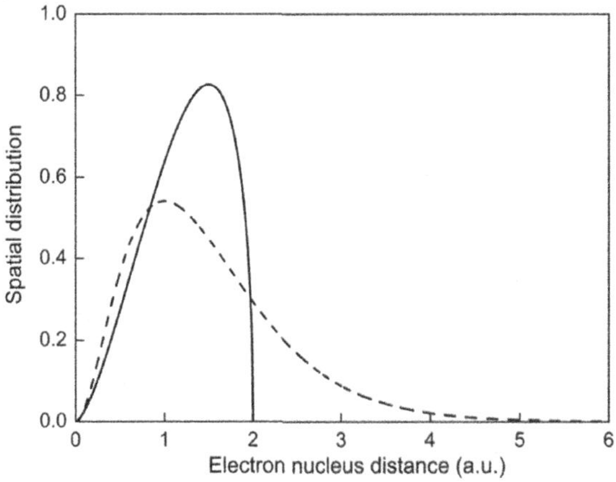

Fig. 2.14 Quantum (dashed curve) and classical (full curve) spatial distribution of hydrogen electron in its fundamental state

Whereas the momentum distribution is exact, the radial distribution

$$\rho_{E_o}(\vec{r_e}) = \frac{16}{\pi r_{e_o}^3} r_e^2 \left(\frac{R_o}{r_e} - 1\right)^{1/2} \tag{2.25}$$

is truncated since the term $\frac{R_o}{r_e} - 1$ ($R_o = \frac{1}{E_o} = 2$a.u.) is necessarily positive. Hence, the maximum of the spatial distribution has a maximum at 2 a.u., as shown in Fig. 2.14, whereas the momentum distribution is not limited in space.

To overcome the strong distortion of classical radial density, many models have been built. However, whatever the approximations (cf. Appendix B), classical distributions are far to successfully represent the reality given by quantum mechanics. Nevertheless, the improvement of cross sections is clear, especially for ionization (cf. Chap. 3).

Another way to obtain the same distributions is to use Kepler's equations, assuming that orbits are not necessarily circular [20]. The method requires for each randomly selected orbit to solve the following equation

$$\theta = \alpha - e\sin\alpha 0 \tag{2.26}$$

where α and e are the eccentric angle and the eccentricity of the orbit, respectively. In expression (2.26), θ is randomly chosen between 0 and 2π, and e^2 is randomly selected between 0 and 1.

With the knowledge of α, the distances and momenta are easily determined. In the case of H atom,

$$\begin{aligned} r_e &= 1 - e\cos\alpha \\ p_e &= \frac{\left(1 - (e\cos\alpha)^2\right)^{1/2}}{1 - e\cos\alpha} \end{aligned} \tag{2.27}$$

2.3.5 Integration Method and Variable Time Step

Among all the iterative integration methods (Picard, Taylor as a few examples), Runge–Kutta numerical method of order 4 is the most utilized one in the treatment of multi-body classical collisions. To describe this method, we only describe the second order method. Suppose time t as the variable, and a point M_n with coordinates $(t_n, y(t_n))$ at iteration n. Rather than calculating the coordinates $(t_{n+1}, y(t_{n+1}))$ of the next point M_{n+1}, the Runge–Kutta method consists of determining an intermediate point M_{int}. The procedure is the following (Fig. 2.14). Point B is the crossing point between slope $(dy/dt)_{t_n}$ at M_n and the vertical line at $t = t_{n+1}$. At time $t = t_n + (t_{n+1} - t_n)/2$ (corresponding to point C), the slope is calculated again. The same slope is used from M_n to determine D, which is much closer to M_{n+1} than B.

At order 4, four coefficients k_i are first calculated, given by

$$\begin{aligned} k_1 &= \Delta t f(y_n, t_n) \\ k_2 &= \Delta t f\left(y_n + \frac{k_1}{2}, t_n + \frac{\Delta t}{2}\right) \\ k_3 &= \Delta t f\left(y_n + \frac{k_2}{2}, t_n + \frac{\Delta t}{2}\right) \\ k_4 &= \Delta t f\left(y_n + k_3, t_n + \frac{\Delta t}{3}\right) \end{aligned} \tag{2.28}$$

Thus the next point is determined using the following relation

$$y(t_{n+1}) = y(t_n) + \frac{1}{3}\left(\frac{k_1}{2} + k_2 + k_3 + \frac{k_4}{2}\right) \tag{2.29}$$

The problem is to determine the time step Δt. In a collision, the projectile interacts with the target, especially at small internuclear distances, at which the trajectory curve is small. At these distances, Δt has to be small, whereas, in the region where the interaction is small, the trajectory is quasi linear, so that Δt can be large. Finally, Δt depends onto the intensities of the potentials involved during the collision. To take into account the potential variations, i.e. the Δt variations, an expression of Δt was formulated to treat $H^+ + H$ collisions [20].

$$\Delta t^2 = \frac{\varepsilon^2}{\frac{v^2}{r^2} + \frac{\varepsilon}{2r}\frac{dv}{dr}} \tag{2.30}$$

where v and r are relative velocity and distance, respectively, and ε is a parameter that depends onto the collision to analyze. In most of the problems, ε was fixed to 0.5. This value is sufficiently small for three-body collisions. However, a value of 0.1 gives better precision when at least four body are involved during the collision (see Appendix C for more details).

Expression (2.30) is easily generalized to a N-body problem [20].

$$\Delta t^2 = \frac{\varepsilon^2}{\sum_{i<j} \frac{p_{ij}^2}{\mu_{ij}^2 r_{ij}^2} + 0.5\varepsilon\left(\sum_{i<j} \frac{1}{\mu_{ij}^2} \frac{dV_{ij}^2}{dr_{ij}}\right)^{1/2}} \tag{2.31}$$

In expression (2.31), p_{ij}, r_{ij}, μ_{ij} and V_{ij} are the relative momenta and distances, the reduced mass, and the potential between particles i and j, respectively.

2.4 Interaction Potentials

2.4.1 Coulombic Potentials

Despite their simplicity, Coulombic potentials are the most useful one to treat a collision. They are mainly utilized for electrons interacting with a positive charge, and to describe interaction between two positive charges, i.e. nuclei. Unfortunately, Coulombic potentials have strong limits.

- Firstly, when highly bound electrons are involved, the potential V is so large and the distance between the electron and the target so small that time step would have to be very small. Consequently the resolution of motion equations would be time consuming. As seen in Fig. 2.15, calculated time step strongly increases when increasing the distance r_e. For H, Ne and Ar atoms, collision times for Ne and Ar are about 100 times and 400 times larger than that for H, it is clear that the treatment of 1s electrons for heavy target is not reasonable.

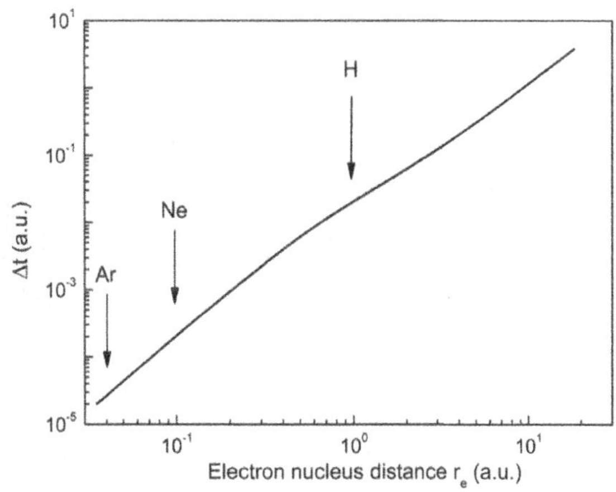

Fig. 2.15 Average calculated time step to describe the trajectory of a 1 s electron as a function of the distance r_e between the electron and the nucleus. The arrows indicate the distances r_e for H,Ne and Ar

- Secondly, the number of electrons may be so high that a complete treatment is unlikely to be made. Since only single processes, such as single electron capture or single ionization are the most important processes to treat, the easiest method is to consider each electron in a certain state as independent. For example, it is not necessary to consider the 8 2*l* electrons of Neon target, since each electron has the same energy and contribute equally to single ionization or electron capture in a collision.
- Finally, electron–electron interaction cannot be described using the only Coulombic potentials. As an example, consider the He atom, with two electrons on a 1 s state. Figures 2.16 and 2.17 represent typical trajectories of *He* electrons with random initial positions and velocities, using the correct total energy of –2.9 a.u.

One of the electrons remains bound on the target nucleus (Fig. 2.16), while the second electron, after a certain time, is ejected and becomes free (Fig. 2.17), due to electron correlation. Consequently, the He atom is not stable. This is due to the fact that certain distances between both electrons are so small that interaction between them is much stronger than electron-nucleus interaction, inducing a spontaneous ejection of one electron, i.e. autoionization. This can also be seen in terms of electron energies (Fig. 2.18). At integration times of the order of 150 a.u., the electron energies drastically change, giving rise to a positive energy for one electron (top of Fig. 2.18).

To circumvent this problem, the first simplest idea would be to initially put both electrons as far as possible from each other. It requires that one electron position is symmetric to the other electron position nucleus, and that their momentum vectors are opposite, whereas He^{2+} is at rest.

In principle, this is the ideal case to maintain the He atom stable. However, due to fluctuations that are specific to computation errors, the He atom is no more stable

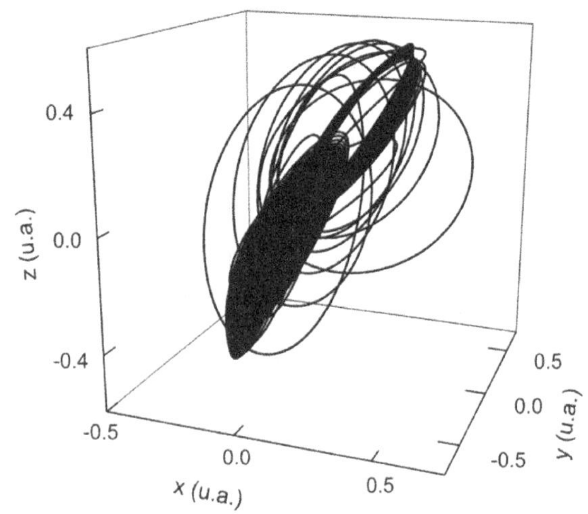

Fig. 2.16 Trajectory of one electron of He, remaining bound during integration time

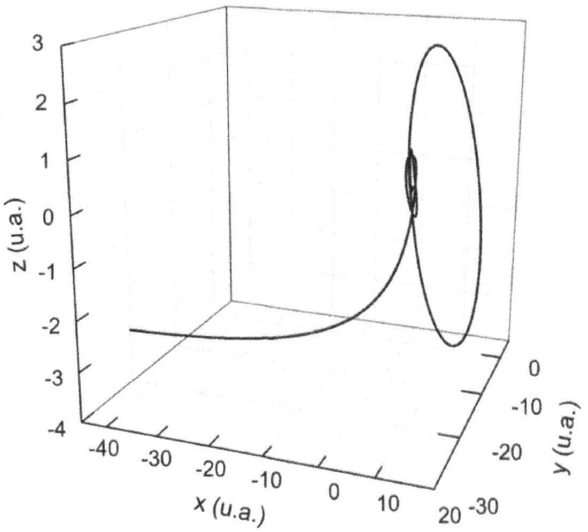

Fig. 2.17 Trajectory of the second electron of He. After a certain time, due to electron correlation, the electron is ionized

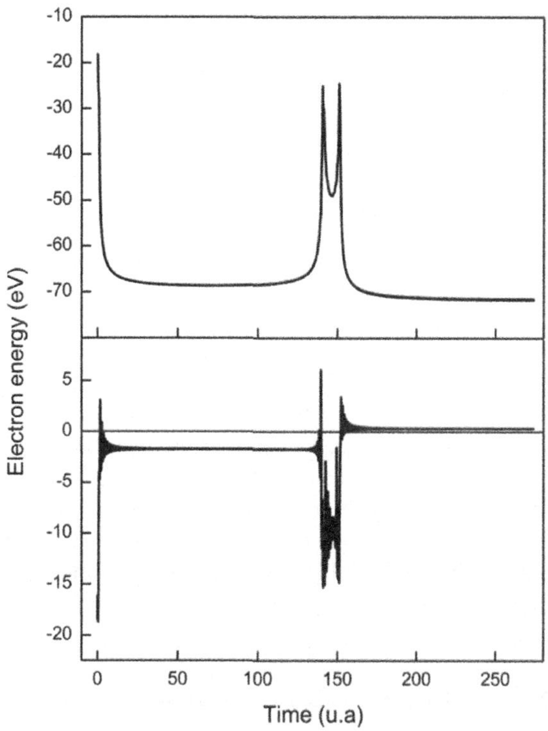

Fig. 2.18 Evolution of He electron energies as a function of integration time. Due to electron correlation, instabilities occur at t of the order of 150 a.u., giving rise to the ejection of one electron

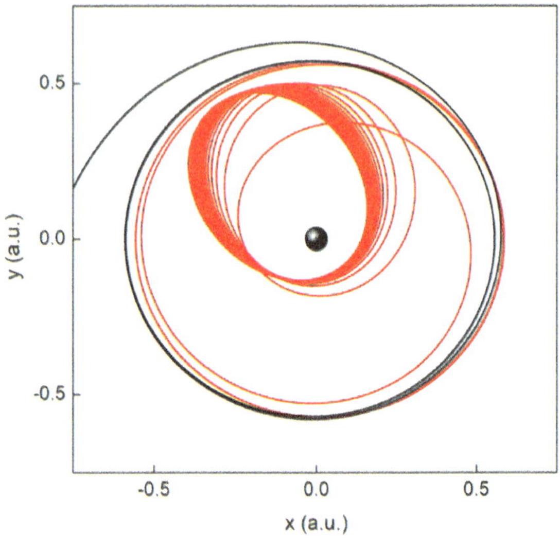

Fig. 2.19 Electron trajectories (full black and red curves) around He nucleus with symmetric initial conditions in position and momentum

after a certain time (Fig. 2.19). Thus, one will have to introduce additional potentials to make it stable.

2.4.2 *Effective Coulombic Potentials*

2.4.2.1 Principle and Motivation

When several electrons are involved, electron correlation plays a decisive role in many cases. However, it is well-known that taking into account electron correlation is a very difficult task. For this reason, independent electron models are used. Electron correlation is neglected, but the charge as viewed by an active electron is not the charge Z of the nucleus, but an effective charge that depends on the number of electrons on the target (or/and on the projectile) and the relative distance of this active electron. Many models were introduced to calculate effective potentials, depending onto the problem to be considered. In the following, we will review the different models that have been created to treat multi electron processes.

2.4.2.2 Particular Case of He Atom

Helium atom consists of a nucleus of charge 2 and two electrons on the same orbital 1s. Both have the same energy, but not the same ionization potential. The first ionization potential is 0.904 a.u. while the second one is 2 a.u. One choice for the potential is to write the total energy as to take into account the screening effect due to the passive electron. Consequently, the potentials are $-Z/r$ and $-(Z-1)/r$. The problem is that $E = -2.5$ a.u., which is far from the expected value of -2.904 a.u.

$$E = \frac{-Z^2}{2} - \frac{-(Z-1)^2}{2} \tag{2.32}$$

Another method is to write

$$E = -2\frac{{Z_{eff}}^2}{2} \tag{2.33}$$

where $Z_{eff} = 1.704$ is the effective charge calculated to obtain the correct total energy. This value for Z_{eff} is close to other values using more sophisticated models [21].

2.4.2.3 Zener and Slater Potentials

In 1930, simple rules have been set up giving approximate analytic wave functions for a given atom or ion, in order to determine the effective charge Z_{eff}. First analytical expressions of wave functions were written for atoms of atomic numbers Z between 4 and 10 [22]. Then, a simple formula for Z_{eff} was deuced in which $n^* = 2$, n is the number of electrons in the L shell and the quantities ϵ_i are the successive ionization potentials of the L shell.

$$Z_{eff} = n^* \sqrt{\frac{\sum_i \epsilon_i}{n}} \tag{2.34}$$

In analogy with the method of Zener, simple rules were set up by Slater for all the atoms in any stage of ionization [23]. The shielding constants and other parameters were not fixed by variation methods but are rather adjusted to get agreement with empirical values.

In the present model, the nodes in the wave functions are, like in Zener's model, considered to be unimportant, so that they are neglected, and only the radial part of the wave function is retained

$$\psi(r) = r^{n^*-1} e^{-\frac{Z-\sigma}{n^* r}} \tag{2.35}$$

where n^* is an effective quantum number that can be real, and σ a screening constant, giving rise to an effective charge $Z_{eff} = Z - \sigma$. The values of n^* and σ are reported in Tables 2.1 and 2.2, respectively.

Table 2.1 Effective quantum numbers n^* as a function of the principal quantum number n

n	1	2	3	4	5	6
n^*	1	2	3	3.7	4	4.2

Table 2.2 Screening constant values due to electron i on electron j

Orbital of electron j	1s	2l	3s, 3p	3d, 3f	4s, 4p	4d	4f	5s, 5p
1s	0.31	0	0	0	0	0	0	0
2l	0.85	0.35	0	0	0	0	0	0
3s, 3p	1	0.85	0.35	0	0	0	0	0
3d, 3f	1	1	1	0.35	0	0	0	0
4s, 4p	1	1	0.85	0.85	0.35	0	0	0
4d	1	1	1	1	1	0.35	0	0
4f	1	1	1	1	1	1	1	0.35
5s, 5p	1	1	1	1	0.85	0.85	0.85	0.85

2.4.2.4 Morse Potentials

A few years later, an empirical formula was determined for diatomic molecules. From Schrödinger equation

$$\Delta\psi + \frac{8\pi^2\mu}{h^2}[W - E(r)]\psi = 0 \tag{2.36}$$

where W describes the vibrational energy levels and $E(r)$ the combination of the repulsion energy and the electron energy, the author proposed for $E(r)$, which is a priori a complicated function of r and the electronic quantum numbers, a simple formulation, introducing a small number of free parameters, depending onto the nature of the molecule and the level considered

$$E(r) = A + De^{-2a(r-r_o)} - 2De^{-a(r-r_o)} \tag{2.37}$$

The parameters A, D, a and r_o are found, according to boundary conditions and other properties that are not explicated here, using the relations

$$\begin{aligned} D &= \frac{\omega_o}{4x} \\ a &= 0.2454\sqrt{\mu\omega_o x} \\ r_o^3\omega_o &= C \end{aligned} \tag{2.38}$$

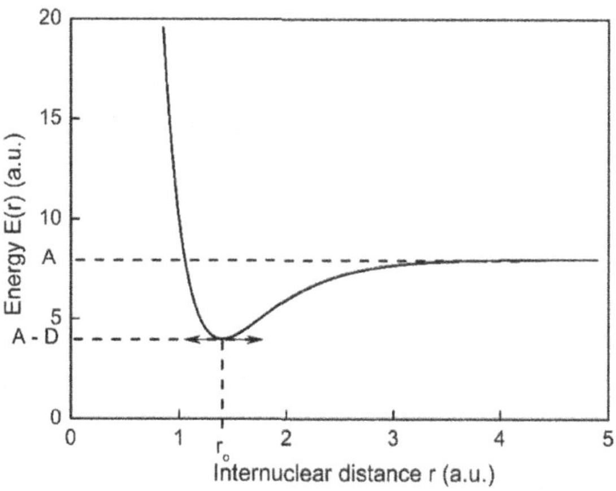

Fig. 2.20 General shape of a Morse potential. At r_o, the potential is minimum, and its value is $A - D$. The dissociation energy, when internuclear distance tends towards infinity, is A

The quantity ω_o has the dimension of an angular velocity, x is a constant that takes place in the expression of W, A is a constant and μ is the reduced mass of the two nuclei.

All the constants were tabulated for quasi symmetric molecules. To apply formula 2.37 to very asymmetric molecules, a factor has to be added.

The general shape for $E(r)$ is shown in Fig. 2.20. When r tends toward 0, $E(r)$ is high but does not tend towards infinity. It is easy to show that $E(r)$ is minimum for r_o, which is the equilibrium distance.

Based on Morse potentials, many other potentials were created in order to improve the agreement with other calculations. These models were built on the same following properties:

- Any acceptable form for the potential energy should have a large value as the nuclei come together.
- It should pass through a minimum at the equilibrium separation.
- It should approach the dissociation energy as the nuclei come apart.

Many functions have these properties, their form is flexible, but each of them presents disadvantages ([24] and references therein). The main disadvantage is the number of parameters involved in potential expressions. Morse or Manning and Rosen [24] have the advantage to have only three parameters, but these functions are generally not adapted for molecules because of their lack of flexibility and precision, especially between r_o and infinity. Other functions are more flexible, since they introduce six or seven parameters [24], but some of the parameters are not easy to determine.

Whatever the form of the potential, the corrections brought to the Morse potential by other authors are significant. However, there does not seem to be any universal forms for a potential energy function since each molecule have their own properties.

In contrast with chemistry, for which high precision in the calculations is required, Morse potentials are sufficient to treat atomic collisions. Indeed, it is well-known that uncertainties of the order of 20% in the determination of total, differential, or partial cross sections, are reasonable in most cases. This remark is true when a few body problem has to be solved; However, for multi electronic targets such as *Ne* or Ar, Morse potentials have to be replaced by more sophisticated potentials.

2.4.2.5 Nonhydrogenic One-Electron Models

For many electron targets, it is often adequate to treat the problem within a one-electron model and employ the independent electron approximation for atomic shell structure. For an accurate calculation, an interaction potential that takes into account the screening of the target nucleus by the electrons is needed. After several theoretical works to parametrize potentials, an expression of $V(r)$ for one electron could be formulated [25]

$$V(r) = -\frac{2}{r}[(Z-1)\Omega(r)+1] \tag{2.39}$$

with

$$\Omega(r) = \frac{1}{\alpha d(Z-1)^{0.4}\left(e^{\frac{r}{d}}-1\right)+1} \tag{2.40}$$

In expressions (2.39) and (2.40), Z is the atomic number of the atom, and α and d are to be determined by adjustment of the bound energies. The average value of α and d are of the order of 1 and 0.8, respectively.

The results obtained using this model, also called GSZ model from the name of their authors, compared favorably with those obtained using Hartree–Fock-Slater (HFS) wave functions.

A few years later, this work was extended to ionized targets [26, 27]. A complete formulation of potentials was determined for all atoms and ions with Z smaller than 54. However, despite all the efforts made, somewhat disappointing was the fact that the evaluation of the energies and parameters gives only on a reasonable good correspondence with other more precise evaluations. Nevertheless, the results are useful for a large variety of collision problems that involve target atoms in their fundamental and ionized states. Note that these potentials can also be used for the representation of partially-stripped projectile ions, especially at high energies, where target and/or projectile ionization occur.

To extrapolate for larger values of Z, it is necessary to base the work upon a relativistic HFS model, which is not of our purpose in the present book.

2.4.2.6 One-Electron-Model for Atomic Collisions

In all the models described above, the electrons are supposed to be static, i.e. on a stable orbit around the nucleus. In fact, during a collision, a target electron is dynamic since it is likely to change its trajectory, as it is perturbed by the fields created by the target and the projectile.

As seen in Fig. 2.21, when an active $3l$ Ar electron is ejected from its initial trajectory, the charge seen by this electron strongly depends on its distance r_e with the nucleus. At large distances, the effective charge is 1, giving rise to a potential $V(r_e) = -1/r_e$. On the contrary, at very small distances, the 1s and $2l$ electron do not screen anymore the nucleus, inducing a potential $V(r_e) = -18/r_e$.

To take into account the evolution of Z with r_e, the potentials $V(r_e)$ were determined from quantum mechanics. The simplest form for $V(r_e)$ is given by

$$V(r_e) = -\frac{Z_o}{r_e} - \left(\frac{Z_1 + Z_2 r_e}{r_e}\right) e^{-\alpha r_e} \tag{2.41}$$

where Z_o, Z_1, Z_2 and α are adjustable parameters. For 3 l state of Ar for example, $Z_o = 1$, $Z_1 = 17$, $Z_2 = 3$ and $\alpha = 2.15$. In fact, in this example (Fig. 2.22), the potential is a sum of several expressions, because the $2l$ electrons also play a role.

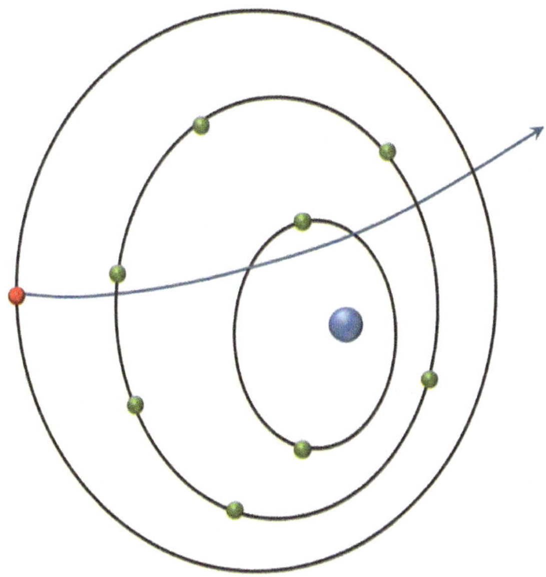

Fig. 2.21 Schematic trajectory of a 3l electron (full red circle), ejected from its initial state and crossing the electronic 2l and 1s electron trajectories (full green circles)

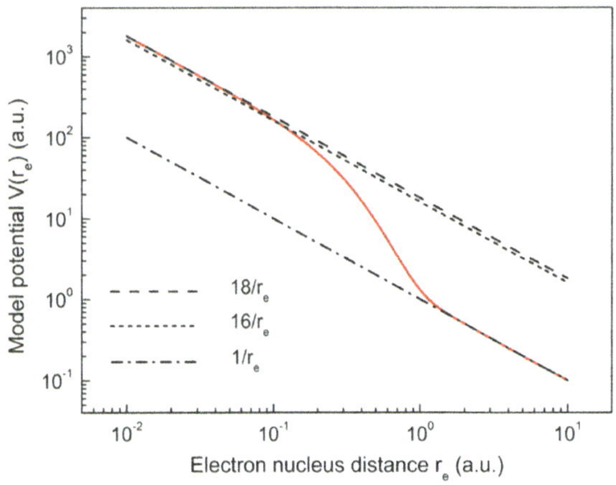

Fig. 2.22 Absolute value of the model potential $V(r_e)$ (red curve) as a function of the distanc er_e between a $3l$ electron and the Ar nucleus, using a similar form than that used in relation (2.41). The straight lines in log–log scale are the asymptotic Coulombic potentials

2.4.3 Inclusion of Electron Correlation

2.4.3.1 Motivation

It has been mentioned above (Sect. 2.4.1) that the introduction of electron correlation for the target leads to spontaneous autoionization. Consequently, since electron correlation plays a decisive role in ionization, electron capture and post collision effects, such as autoionization, electron independent models cannot be used anymore. Thus, the only way to take into account electron correlation is to add a phenomenological potential in order to force the electrons to stay around the nucleus, despite the intensity of electron electron interaction. This kind of potential was first introduced successfully in nuclear physics [28] to mimic the Pauli exclusion principle and was then extended in atomic physics to simulate both Pauli exclusion and Heisenberg uncertainty principles [29].

2.4.3.2 Expression of the Phenomenological Potential

The particular form of the potential V_H and V_P associated with the Heisenberg and Pauli principle, respectively, are

$$V_H(r,p) = \frac{\xi_H^2}{4\mu\alpha_H r^2}\exp\left(\alpha_H\left[1-\left(\frac{rp}{\xi_H}\right)^4\right]\right) \tag{2.42}$$

$$V_P(r,p) = \frac{\xi_P^2}{4\mu\alpha_P r^2}\exp\left(\alpha_P\left[1-\left(\frac{rp}{\xi_P}\right)^4\right]\right) \tag{2.43}$$

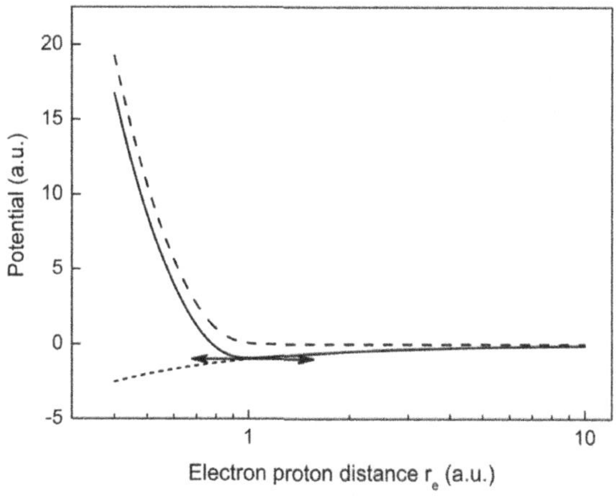

Fig. 2.23 Shape of the potential $-\frac{1}{r_e} + V_H(r_e, p_e)$ as a function of r_e, for $p_e =$ 1 a.u. Dashed curve, $V_H(r_e, p_e)$ short dashed curve $-\frac{1}{r_e}$, Coulombic potential; full curve, sum of both potentials. The arrow indicates the minimum of the total potential

In the above expressions, α_H, α_P, ξ_H and ξ_P have to be determined. In atomic collisions, $V_P(r, p)$ is taken into account only if the number of active electrons is larger than 2. In the following, $V_P(r, p)$ will not be introduced anymore.

Introducing $V_H(r, p)$, the total Hamiltonian now writes

$$H = \sum_i T_i + \sum_{i<j} \frac{q_i q_j}{r_{ij}} + \sum_{i<j} V_H\left(r_{ij}, p_{ij}\right)_H \tag{2.44}$$

The first term is the sum of the kinetic energies, the second the sum of Coulombic potentials. The quantities r_{ij} and p_{ij} are the relative distances and momenta, respectively, between particles i and j.

As an example, for the simple case of atomic hydrogen, relation (2.44) becomes

$$H = \frac{p_H^2}{2M_H} + \frac{p_e^2}{2} - \frac{1}{r_e} + V_H(r_e, p_e) \tag{2.45}$$

The coulombic potential is attractive, while $V_H(r_e, p_e)$ is repulsive, inducing a minimum in distance for a fixed value of p_e in the sum of both, as seen in Fig. 2.23. The presence of the minimum allows the electron to be confined in a certain region of space around the nucleus.

2.4.3.3 Determination of the Phenomenological Potential Parameters

To determine the values of α_H and ξ_H, a process of minimization of the total energy has to be engaged. Nevertheless, rather than applying such method for each atom or molecule, α_H and ξ_H have been fixed to 4 and 0.943. Then the Hamiltonian H has been minimized by performing the best choice for the quantities r_e and p_e. Figure 2.24 presents the results for total energy of atom electrons as a function of

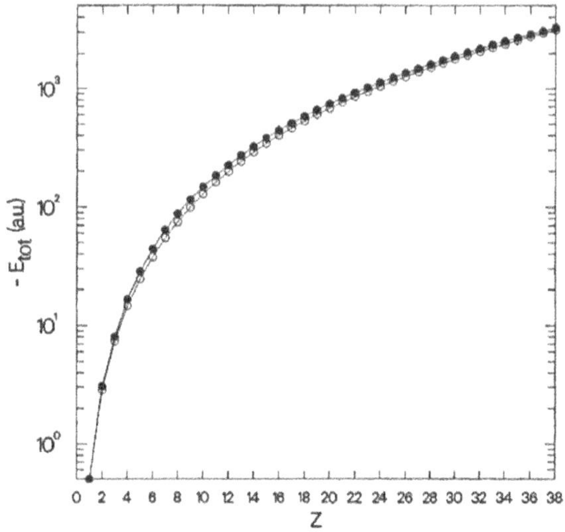

Fig. 2.24 Total ionization energy of electrons in an atom of atomic number Z as a function of Z. Full circles, calculations using $V_H(r_e, p_e)$; empty circles, Hartree–Fock calculations

atomic number Z. The results of the above procedure (full circles) are compared to the total energy calculated using a non-relativistic Hartree-Foch method (empty circles) [30, 31]. The agreement between both calculations is remarkable.

To check the present model, first and second ionization potentials (Figs. 2.24 and 2.25, respectively) were also calculated (full circles) and compared with Hartree–Fock calculations (empty circles). The agreement between both methods is quite satisfactory, even if large differences appear for some specific values of Z. In these

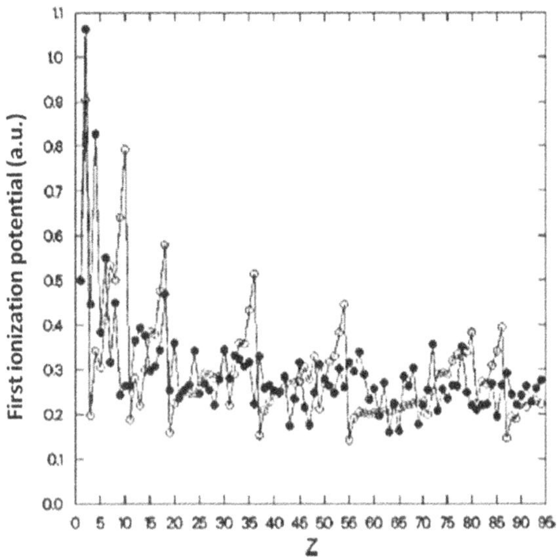

Fig. 2.25 First ionization potential energy as a function of Z. Full circles, calculations using $V_H(r_e, p_e)$; empty circles, Hartree–Fock calculations

particular cases, parameters α_H and ξ_H are adjusted in order to obtain the correct energy values (Fig. 2.26).

The values of r_e and p_e used to perform calculations are summarized in Figs. 2.27 and 2.28 for Z varying from 1 to 38.

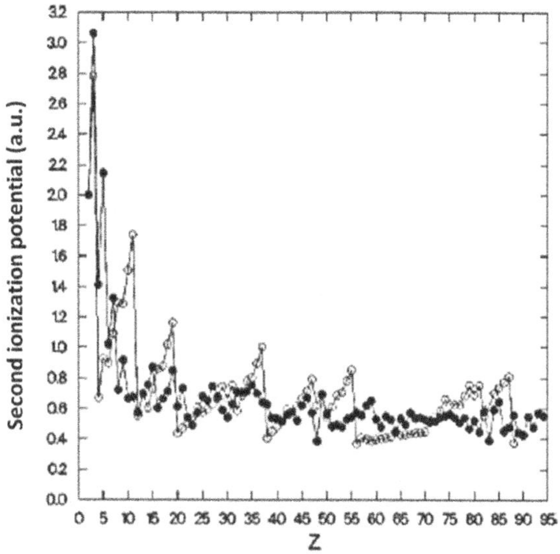

Fig. 2.26 Second ionization potential energy as a function of Z. Full circles, calculations using $V_H(r_e, p_e)$; empty circles, Hartree–Fock calculations

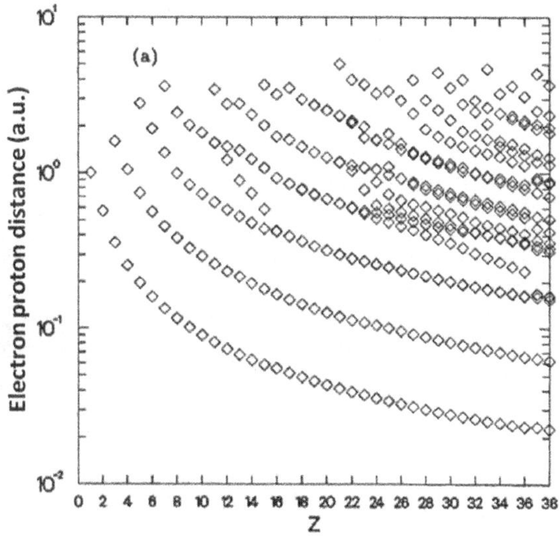

Fig. 2.27 Electron-nucleus distances r_e chosen to calculate ionization energies, by using Coulombic and phenomenological potentials, for Z varying from 1 to 38

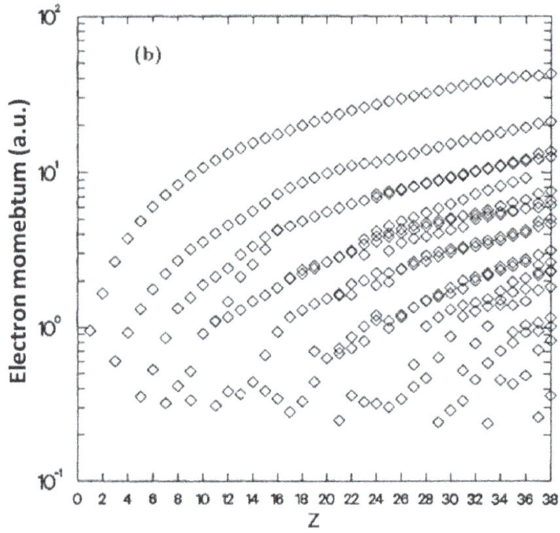

Fig. 2.28 Electron momenta p_e chosen to calculate ionization energies, by using Coulombic and phenomenological potentials, for Z varying from 1 to 38

2.4.3.4 Application to He Atom

Suppose a He atom in which electron-nucleus and electron–electron interactions are considered.

At time $t = 0$, both electron positions and momenta are randomly chosen, and the phenomenological potential V_H is added between each electron and the nucleus. Among these initial conditions, some of them give rise to stable trajectories, as seen in Fig. 2.29. In first approximation, the He atom can be considered as stable. Nevertheless, even if trajectories are continuous around the nucleus, they present

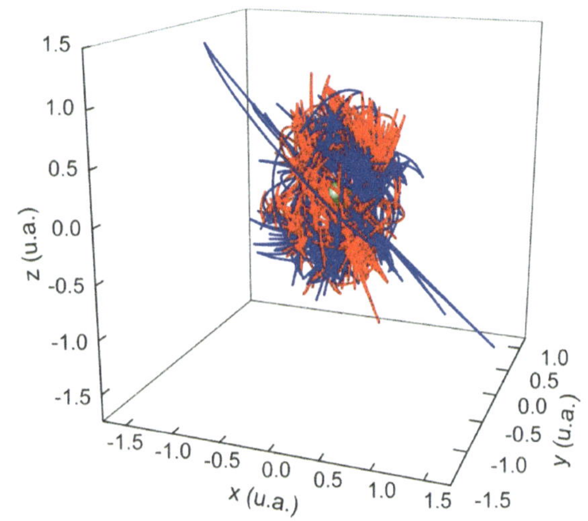

Fig. 2.29 3D plot trajectories of He electrons, submitted to electron-nucleus and electron–electron interactions, using Coulombic and phenomenological potentials

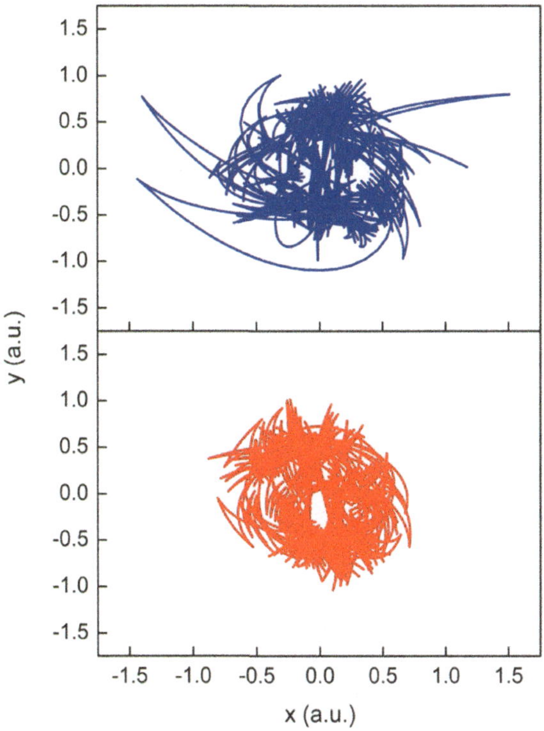

Fig. 2.30 2D plot trajectories of He electrons (blue and red curves), submitted to electron-nucleus and electron–electron interactions, using Coulombic and phenomenological potentials. Strong singularities are observed far from the nucleus. The initial electron positions and momenta are randomly chosen

strong singularities (Fig. 2.30 in two dimensions), so that, at these particular points, the derivatives are not continuous.

It is easy to show that, at these singular points, potential energy is well defined, and that the first derivative is minimum or maximum. At the maximum of potential energy, the second derivative can be negative, meaning that the equilibrium is not stable. Our experience shows that, with such type of initial configurations, ionization cross section diverge as impact parameter increases.

Consequently, if phenomenological potentials are introduced for calculations, only specific initial states have to be introduced. For He atom, it has been shown that the stability is fulfilled (Fig. 2.31) when both electron positions are symmetric to each other with respect to the nucleus position, and when their initial momenta are opposite to each other.

2.4.3.5 Alternatives to Electron Correlation

Between introducing or not the electron correlation, alternatives have been developed in order to partially take into account the interaction between both electrons. Some authors introduced electron–electron interaction before and after the collision itself, whereas this interaction is cut off during the time where the inelastic processes occur.

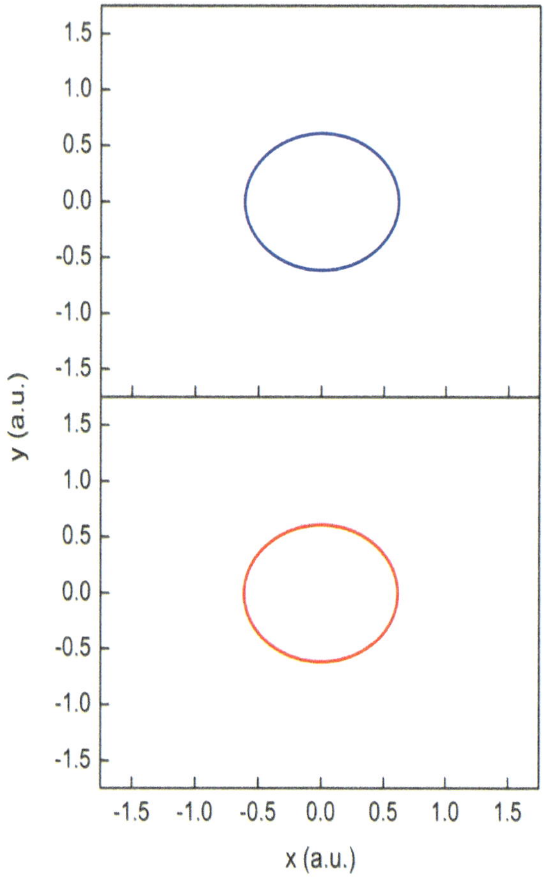

Fig. 2.31 2D plot trajectories of He electrons (blue and red curves), submitted to electron-nucleus and electron–electron interactions, using Coulombic and phenomenological potentials. The electrons are symmetric to each other with respect to the nucleus position, and when their initial momenta are opposite to each other

Another way was to introduce Gaussian density distributions for the electron [32], in order to calculate the strength of electron correlation. A simpler method, which is described below, was the introduction of a dynamic screening of the target nucleus as seen by each of the two active electrons [33].

In the Hamiltonian [33], the electron–electron interaction is replaced by an alternate form $V_{ee}\left(\overrightarrow{r_1}, \overrightarrow{r_2}, \overrightarrow{p_1}, \overrightarrow{p_2}\right)$, where $\overrightarrow{r_i}$ and $\overrightarrow{p_i}$ are the distances between the nucleus and the electron i, and the momentum of this electron, respectively. The term V_{ee} embodies a limited amount of electron correlation. More precisely, only the radial correlation is taken into account in the formula of V_{ee}.

Starting from the wave function

$$\psi(r) = \left(\frac{\zeta^3}{\pi}\right)^{1/2} e^{-\zeta r} \tag{2.46}$$

where ζ is a constant,

the magnitude of the charge $Q(r, \zeta)$ due to a single electron which is found within a sphere of radius r is

$$Q(r, \zeta) = 1 - [1 + 2\zeta r(1 + \zeta r)]e^{-2\zeta r} \tag{2.47}$$

Thus the potential V_{ee} is the sum of two terms V_{ee}^{i} $(i = 1, 2)$

$$V_{ee}^{i}\left(\left|\vec{r_t} - \vec{r_i}\right|, \zeta_j\right) = \frac{1}{\left|\vec{r_t} - \vec{r_i}\right|} - \left[1 - \left(1 + \zeta_j\left|\vec{r_t} - \vec{r_i}\right|\right)\right]e^{-2\zeta_j\left|\vec{r_t} - \vec{r_i}\right|} \tag{2.48}$$

2.5 Collision and Post-Collision Processes

2.5.1 *Collision Processes*

Suppose a multiply charged ion A^{q+} colliding on an atomic or a molecular target B. The different processes that may occur during the collision are summarized below

$$A^{q+} + B \rightarrow A^{q+} + B \tag{2.49a}$$

$$A^{q+} + B \rightarrow A^{q+} + B^{*} \tag{2.49b}$$

$$A^{q+} + B \rightarrow A^{q+} + B^{s+} \tag{2.49c}$$

$$A^{q+} + B \rightarrow A^{(q-r)+} + B^{r+} \tag{2.49d}$$

The most probable process is elastic scattering (ES) (2.49a), where the projectile and target natures are unchanged. The target may be excited (2.49b) or ionized (2.49c) (s electrons are removed from the target and become free). There is also a probability that r electrons are removed from the target and captured by the projectile (2.49d). This process is called charge transfer or electron capture (EC).

Of course, combinations of excitation, ionization and electron capture is possible. The probability for such combination strongly depends onto the projectile velocity v_P compared to the velocity v_e of the active electrons.

- if $v_P \ll v_e$, electron capture is the dominant process.
- if $v_P \gg v_e$, ionization and/or excitation are the most probable processes.
- if $v_P \sim v_e$, i.e. at intermediate velocities, all the processes compete.

Collision processes are efficient on a distance that is similar to the size of the target, typically between a few a.u. for H up to 100 a.u. for a heavy atom such as Xe. Thus, for a typical projectile velocity $v_P = 1$ a.u., collision time varies from ~10^{-16} to 10^{-14} s.

2.5.2 Post-Collision Processes

After the collision, the projectile and/or the target are mainly in excited states. Two kinds of post-collision processes are distinguished.

- The projectile and/or the target deexcite via photon emission (radiative deexcitation) or Auger effect (electron emission).
- In the case of a molecular target, fragmentation may occur.

These processes, at intermediate velocities, take more time to occur than collision processes. Consequently, in first approximation, collision and post-collision processes can be considered as independent, and thus be separated. However, even if this separation is convenient in most cases, we have to keep in mind that radiative and Auger deexcitations follow an exponential law, decreasing in magnitude, so that the latter processes begin during the collision, and not after. On a classical point of view, this means that collision and post-collision processes have to be treated at the same time.

2.5.3 Processes Selection Criteria in Classical Mechanics

2.5.3.1 Introduction

To distinguish the processes introduced above, the simplest criterion is the sign of final electron energy E_e. If E_e is negative, the electron is bound, otherwise it is free. Because an active electron is in the field of the target and the projectile during a collision, processes selection criteria have to be detailed. If we call E_e^P and E_e^T the electron energies relative to the projectile and the target, respectively, that are defined as follows

$$\begin{aligned} E_e^P &= \frac{1}{2}\left(v_e^P\right)^2 - \frac{q_P}{r_e^P} \\ E_e^T &= \frac{1}{2}\left(v_e^T\right)^2 - \frac{q_T}{r_e^T} \end{aligned} \tag{2.50}$$

Table 2.3 Electron energy signs compared to the target (2nd column) and the projectile (3rd column) according to the processes involved during A^{Z+} + H collision

Processes	E_e^T	E_e^P
ES	<0	>0
SE	<0	>0
SC	>0	<0
SI	>0	>0

q_T, q_P, v_e^T and v_e^P are the projectile and target charges and the electron velocities associated with the target and the projectile, respectively, and r_e^P and r_e^T are the distances between the electron and the projectile and target nuclei, respectively.

In order to simplify the discussion, the criteria will be displayed step by step. by increasing the complexity. Criteria tables will be described in the following for $A^{Z+} + H$ (one active electron) and $A^{Z+} + He$ (two active electrons).

2.5.3.2 A^{Z+} + H Collision System

In A^{Z+} + Hcollision, elastic scattering (ES), single electron capture (SC), single excitation (SE) and single ionization (SI) may occur. In the case of ES and SE, the electron stays bounded on the target, thus $E_e^P > 0$ and $E_e^T < 0$. For SC, the electron is bounded on the projectile, so that $E_e^P < 0$ and $E_e^T > 0$. For SI, the electron is free, and both energies are positive (see summary in Table 2.3).

As seen in this table, ES and SE cannot be separated. The sign of the energies is not sufficient to identify the processes. Consequently, another criterion, that will be discussed further on, has to be added (see Sect. 2.5.3.4).

2.5.3.3 A^{Z+} + He Collision System

In addition to one-electron processes described above, multi-electron processes may occur.

In the simplest case of He target (Table 2.4), the number of processes strongly increases, so that the difficulty to separate them is extended. We call $E_e^P(i)$ and $E_e^T(i)$ $(i = 1, 2)$ the electron energies of electron i. In addition to the processes developed in the previous section, combination of these may occur (Table 2.4). If one electron is captured and the other one is excited or ionized, the processes will be named transfer-excitation (TE) or transfer-ionization. If one electron is excited and the other one is ionized, the process will be called excitation-ionization (EI). After double excitation (DE) or double capture (DC), the target or the projectile have a probability to deexcite via Auger effect (also called autoionization). In these cases, the processes will be called autoionizing double excitation (ADE) and autoionizing double capture (ADC).

Table 2.4 Electron energy signs compared to the target (2nd and 4th columns) and the projectile (3rd and 5th columns) according to the processes involved during $A^{Z+} + He$ collision

Processes	$E_e^T(1)$	$E_e^P(1)$	$E_e^T(2)$	$E_e^P(2)$
ES	<0	>0	<0	>0
SE	<0	>0	<0	>0
SC	>0 <0	<0 >0	<0 >0	>0 <0
SI	>0 <0	>0 >0	<0 >0	>0 >0
DE	<0	>0	<0	>0
DC	>0		>0	
TE	>0 <0	<0 >0	<0 >0	>0 <0
TI	>0 <0	<0 >0	<0 >0	>0 <0
EI	>0 <0	>0 >0	<0 >0	>0 >0
ADC	>0 >0	<0 >0	>0 >0	>0 < 0
ADE	>0 <0	>0 >0	<0 <0	>0 >0

In principle, by using only the signs of the energies, it is not possible to distinguish ADC and TI, since they lead to the same final state. Nevertheless, we will see (Sect. 2.5.3.5) that other criteria can be involved to separate them.

2.5.3.4 Classical Classification of Quantum Numbers

In quantum mechanics, it is well known that quantities like energies are quantized. As an example, the electron energy on H depends on the principal quantum number n, which is an integer, via the relation $E_e = -1/2n^2$, where E_e is expressed in a.u. In a collision, where excitation is the process to analyze, the excited level distribution resembles that presented in Fig. 2.32.

In classical mechanics, the energy electron is not quantized. At the end of the collision, it can take all possible values, inducing a continuous distribution (Fig. 2.33). It is thus necessary to define a classical quantum number, which is real and not integer, each quantum number corresponding to a specific range of electron energies.

It was shown previously that the classical principal quantum number n_c associated with a classical energy $E_c = -1/2n_c^2$ satisfies the following inequalities [34, 35]

$$\left[(n-1)\left(n-\frac{1}{2}\right)\right] \leq n_c \leq \left[n\left(n+\frac{1}{2}\right)(n+1)\right] \tag{2.51}$$

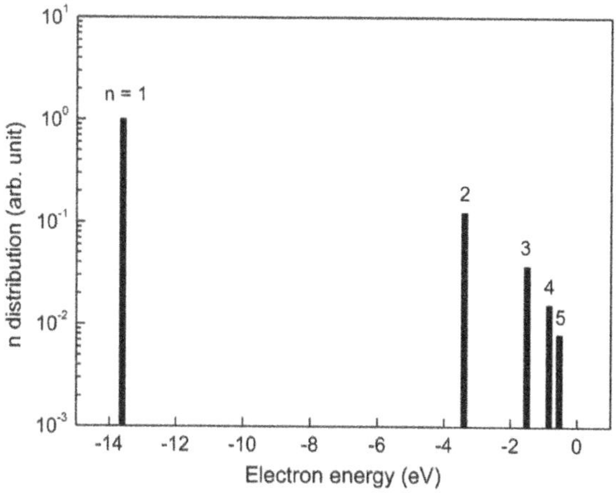

Fig. 2.32 Schematic representation of principal quantum number distribution after excitation of H by a charged ion, as a function of the electron energy (eV)

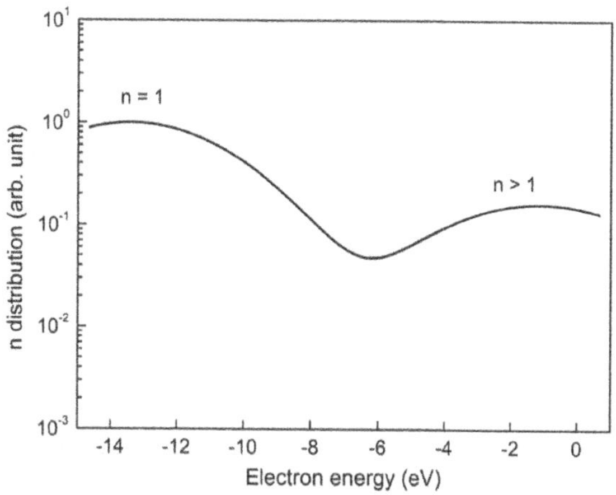

Fig. 2.33 Schematic representation of principal classical number distribution after excitation of H by a charged ion, as a function of the electron energy (eV)

This relation is well adapted when the ratio q/Z, where q is the projectile charge, and Z its atomic number, is close to 1. If it is not so, i.e. for small ratios q/Z, the relation (2.51) can be applied with certain conditions [36]. Using relation (2.51), the energy range corresponding to each quantum number is easily deduced.

A similar analysis was made for angular momentum distributions, by calculating classical angular momentum, defined by

$$\ell_c = \left[\left(x\frac{dy}{dt} - y\frac{dx}{dt}\right)^2 + \left(y\frac{dz}{dt} - z\frac{dy}{dt}\right)^2 + \left(z\frac{dx}{dt} - x\frac{dy}{dt}\right)^2\right] \tag{2.52}$$

The classical angular momentum satisfies the relation

$$\ell \leq \frac{n}{n_c}\ell_c \leq \ell + 1 \tag{2.53}$$

2.5.3.5 Auger Effect or Autoionization

Auger effect may occur after DE of the target, or after DC onto the projectile. As mentioned in Sect. 2.5.3.3, ADC and TI lead to the same final state, since one electron is ejected while the other one is captured by the projectile. Since Auger effect is a pure quantum effect, it is a priori impossible to treat it classically. Nevertheless, some intuitive considerations and criteria can be discussed to try to separate, without any ambiguity, ADC and TI processes.

Probability criteria

At low projectile energy, capture process is dominant, and TI is unlikely to occur. Therefore, if we observe, after a collision, a free electron and a captured electron, the probability for ADC is much larger than that for TI.

In contrast, double capture cross sections are much smaller than TI cross sections at large projectile velocities. Finally, the problem arises at intermediate energies, where both TI and ADC processes compete with similar probabilities. Consequently, the probability criterion is not exempt of difficulties.

Angular criteria

First, we introduce two angles. Since a particle is characterized by its position and velocity, we call θ_p the angle between an axis (Ox) and the vector $\overrightarrow{OM}$ and θ_v the angle between the axis (Ox) and the velocity vector $\vec{v}(M)$. At the beginning of the collision, θ_p and θ_v are far to be equal. At large distances, both angles are close to each other.

In the case of ADE and EI processes, the electron is ejected from the target, whereas the second one remains on it. The ejection occurs from a point close to the origin O of the coordinate frame. Since at large distances $\theta_p \sim \theta_v$ for both processes, the angle criterion does not allow the separation of the processes.

In the case of ADC and TI, the electron is ejected either from the target or from the projectile. Therefore, in most situations, both angles strongly differ.

Time criteria

Even if TE and ADC lead to the same final states, time dependences of such processes are very different (Fig. 2.34). Since SC and SI occur during the collision, the probability for such processes to reach their maximum at time t_o of the order of about

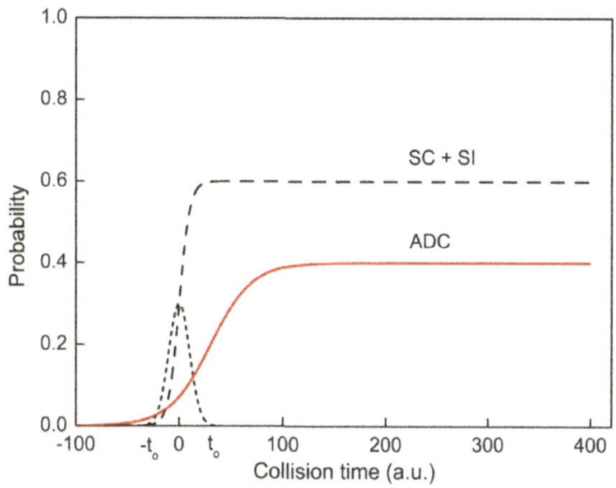

Fig. 2.34 Schematic expected behavior of SI and SC probabilities (dashed curve) and ADC probability (full red curve) as a function of integration time. The short-dashed curve shows a typical time interval during which a quasi-molecule is formed. In this interval, the electrons are shared between the two nuclei

a few tens of a.u. (dashed curve on Fig. 2.34). In contrast, for ADC, since relaxation time is much larger than collision time, the shape of the probability for ADC strongly differs from that for collision processes, since it increases slowly with time. This schematic comparison clearly shows that it is possible, in principle, to find a signature of autoionization by analyzing the collision at various integration times [37].

2.6 Total, Partial and Differential Cross Sections

2.6.1 *Total Cross Sections*

Once a significant number of trajectories have been calculated, total cross sections can be calculated for each process X (ionization, excitation, charge transfer). In the first calculations, the following formula was used

$$\sigma_X = \frac{N_X}{N_{tot}} \pi b_{max}^2 \tag{2.52}$$

In this relation b_{max} is the maximum impact parameter involved during process X, and the quantity N_X/N_{tot} is the ratio between the number of collisions satisfying the criteria for reaction X and the total number of calculated trajectories.

The standard deviation for σ_X is given by

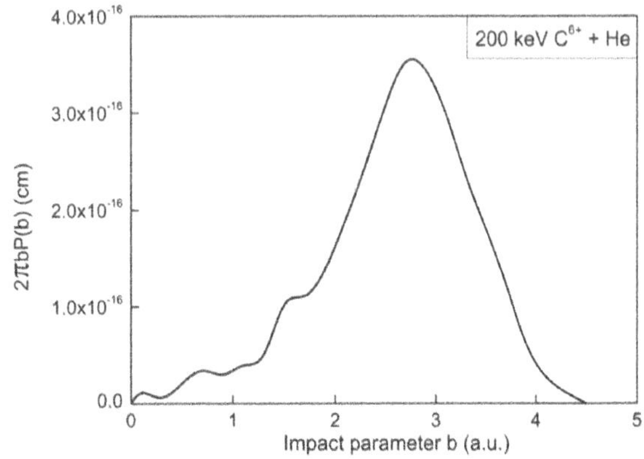

Fig. 2.35 Impact parameter distribution after single capture in 200 keV C^{6+} + He, calculated by means of CTMC method

$$\Delta\sigma_X = \sigma_X \left(\frac{N_{tot} - N_X}{N_{tot} N_X} \right)^{1/2} \tag{2.55}$$

Another method consists in considering impact parameter distributions that are available using classical mechanics. Since all target orientations are possible, the differential cross section $d\sigma_X/db$ in impact parameter b is constant inside a ring whose radius is between b and $b + db$. Thus σ_X is obtained by integration of $d\sigma_X/db$ over b

$$\sigma_X = 2\pi \int_0^{b_{max}} b \frac{N_X(b)}{N_{tot}(b)} db \tag{2.56}$$

To illustrate, we consider the collision C^{6+} + He at 200 keV. Figure 2.35 shows the differential cross section $2\pi b P(b)$, where $P(b) = N_X(b)/N_{tot}(b)$ is the SC probability at each value of b. By integration, we find $\sigma_X = 5.8\ 10^{-16}$ cm^2, whereas, using relations (2.54) and (2.55), $\sigma_X = 5.1 \pm 0.2\ 10^{-16}$ cm^2.

2.6.2 *Partial Cross Sections*

We consider the case of a capture of an electron in an excited state of the projectile. The total cross section σ_C writes

$$\sigma_C = \sum_{n=1}^{n_{max}} \sigma_n \tag{2.57}$$

where σ_n is the partial capture cross section on a state of principal classical number n.

In a similar way, in the knowledge that, for a fixed n, the angular momentum varies from 0 to l_{max}, σ_n is the sum of all partial cross sections σ_{nl} in l.

$$\sigma_n = \sum_{l=1}^{l_{max}} \sigma_{nl} \tag{2.58}$$

2.6.3 *Differential Cross Sections*

Total cross sections are the first step in the understanding of a collision. Even if calculated cross sections coincide with experiment, it does not validate the theory. To confirm the agreement, it is necessary to go into deeper details. As an example, in the case of SI, the electron is ejected in specific directions, with certain probabilities. Moreover, for a given ejection (or detection) angle θ_d, electron energies follow distributions that are far to be uniform. Thus, many experiments and calculations were conducted to determine multi-differential cross sections defined as $d\sigma^{2n}/\left(d\theta_d^1 \ldots d\theta_d^n dE_e^1 \ldots dE_e^n\right)$, where E_e^i and θ_d^i are the energy and the detection angle, with respect to the projectile initial velocity, of electron i, respectively. These detailed quantities are of crucial importance because they give insights on the interactions involved during the collision.

In the particular case of one ejected electron (SI), the quantities to determine are $d^2\sigma_{SI}/(d\theta_d dE)$ and $d^2\sigma_{SI}/(d\Omega_d dE)$ where $d\Omega_d = 2\pi \sin\theta_d d\theta_d$ is the solid angle as seen by the detector. The latter differential cross sections are related to the total cross section by

$$\sigma_{SI} = \int_0^{\pi} \frac{d^2\sigma_{SI}}{d\Omega_d dE} d\Omega_d dE = \int_0^{\pi} \frac{d^2\sigma_{SI}}{d\theta_d dE} d\theta_d dE \tag{2.59}$$

As an example, we consider the collision system $C^{6+} + H$ at 5 meV. Total SI cross sections is 1.2 10^{-15} cm^2. Figures 2.36 shows typical differential cross sections in angle (top of Fig. 2.36) and in solid angle (bottom of Fig. 2.36). Note that the second distribution is obtained by dividing the first one by $2\pi \sin\theta_d$, and that integration of both has to lead to the total cross section.

Figure 2.37 presents typical energy distribution of ejected electrons, integrated over all ejection angles, as a function of the ejected electron energy. The integration of this distribution leads to the total SI cross section.

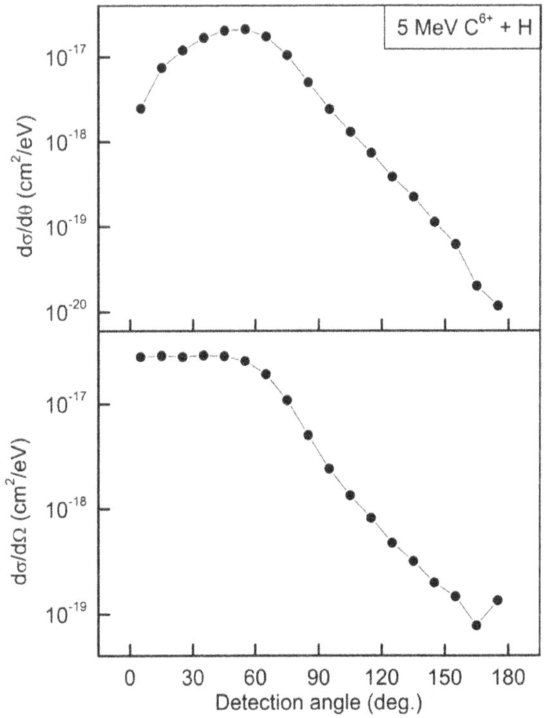

Fig. 2.36 Typical angular distributions following single ionization in 5 meV $C^{6+} + H$ as a function of the detection angle, calculated by means of a CTMC model. Top of the figure, $d^2\sigma_{SI}/(d\theta_d dE)$; bottom of the figure, $d^2\sigma_{SI}/(d\Omega_d dE)$

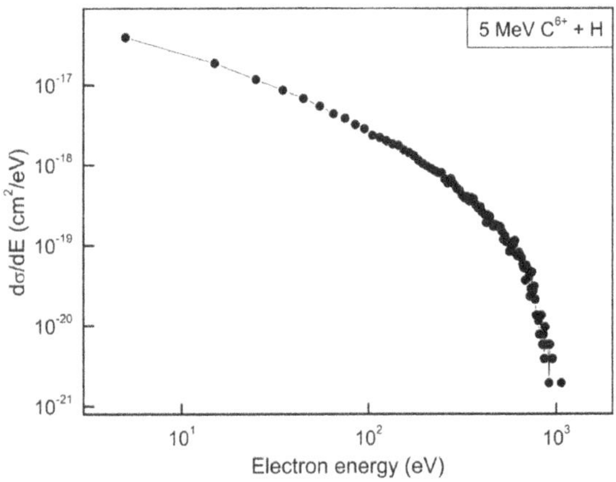

Fig. 2.37 Typical energy distribution of ejected electrons following single ionization in 5 meV $C^{6+} + H$ as a function of the detection angle, calculated by means of a CTMC model

References

1. Gryzinski, M.: Classical theory of electronic and ionic inelastic collisions. Phys. Rev. **115**, 374 (1959)
2. Boyd, R.L.F., Green, G.W.: Electron ionization cross sections using chopped beams. Proc.

Phys. Soc. (London) **71**, 357 (1958)
3. Vriens, L.: Binary-encounter electron-atom collision theory. Phys. Rev. **141**, 88 (1966)
4. Kim, Y.K., Rudd, E.: Binary-encounter-dipole model for electron impact ionization. Phys. Rev. A **50**, 3954 (1994)
5. Bethe, H.: Zur Theorie des Durchgangs schneller Korpuskularstrahlen durch Materie. Ann. Phys. (Leipzig) **5**, 325 (1930)
6. Younger, S.M.: Electron impact ionization cross sections and rates for highly ionized atoms. J. Quant. Spectrosc Radiat. Transfer **26**, 329 (1981)
7. Shah, M.B., Elliott, D.S., Gilbody, H.B.: Pulsed crossed-beam study of the ionisation of atomic hydrogen by electron impact, J. Phys. B **20**, 3501 (1987)
8. Djuric, N.L., Cadez, I.M., Kurepa, M.V.: H_2O and D_2O total ionization cross-sections by electron impact. Int. J. Mass. Spectrom. Ion. Proc. **83**, R7 (1988)
9. Bolorizadeh, M.A., Rudd, M.E.: Angular and energy dependence of cross sections for ejection of electrons from water vapor. I. 50–2000-eV electron impact. Phys. Rev. A **33**, 882 (1986)
10. Schutten, J., DeHerr, F.J., Moustafa, H.R., Boerbom, A.J.H., Kistemaker J.: Gross- and partial-ionization cross sections for electrons on water vapor in the energy range 0.1–20 keV. J. Chem. Phys. **44**, 3924 (1966)
11. Ryufuku, H., Sasaki, K.: Oscillator behavior of charge transfer cross sections as a function of the charge of projectiles in low-energy collisions. Phys. Rev. A **21**, 745 (1980)
12. Bayfield, J.E., Koch, P.M., Gardner, L.D., Sellin, I.A., Pegg, D.J., Peterson, R.S., Crandall, D.H.: In contributed papers if the international conference on atomic physics. In: Marrus, R., Prior, M.H., Shugart, H.A., Berkley (eds.), p. 126 (1978)
13. Kim, H.J., Hvelplund, P., Meyer, F.W., Phaneuf, R.A., Stelson, P.H., Bottcher, C.: Phys. Rev. Lett. **40**, 1636 (1978)
14. Barany, A., Astner, G., Cederquist, H., Danared, H., Huldt, S., Hvelplund, P., Johnson, A., Knudsen, H., Liljeby, L., Rensfelt, K.-G.: Absolute cross sections for multi-electron processes in low energy Arq+–Ar collisions: comparison with theory. Nucl. Instr. Meth. Phys. Res. B **9**, 397 (1985)
15. Niehaus, A.: A classical model for multiple-electron capture in slow collisions of highly charged ions with atoms. J. Phys. B **19**, 2925 (1986)
16. Nielsen, E.H., Andersen, L.H., Barany, A., Cederquist, H., Hvelplund, P., Knudsen, H., MacAdam, K.B., Sorensen, J.: J. Phys. B **17**, L139 (1984)
17. Hirschfelder, J., Eyring, H., Topley, B.: Reactions involving hydrogen molecules and atoms. J. Chem. Phys. **4**, 170 (1936)
18. Wall, F.T., Hiller, L.A., Mazur, J.: Statistical computation of reaction probabilities. J. Chem. Phys. **35**, 1284 (1961)
19. Karplus, M., Porter, R.N., Sharma, R.D.: Exchange reactions with activation energy. I. Simple barrier potential for (H, H_2), J. Chem. Phys. **43**, 3259 (1965)
20. Abrines, R., Percival, I.C.: Classical theory of charge transfer and ionization of hydrogen atoms by protons. Proc. Phys. Soc. **88**, 861 (1966)
21. Olson, R.E.: Electron capture and impact ionization cross sections for multiply charged ions colliding with helium. Phys. Rev. A **18**, 2464 (1978)
22. Slater, J.C.: Atomic shielding constants. Phys. Rev. **36**, 57 (1930)
23. Zener, C.: Analytic atomic wave functions. Phys. Rev. **36**, 51 (1930)
24. Hulburt, H.M., Hirschfelder, J.O.: Potentiel energy functions for diatomic molecules. J. Chem. Phys. **9**, 61 (1941)
25. Green, A.E.S., Sellin, D.L., Zachor, A.S.: Analytic independent-particle model for atoms. Phys. Rev. **184**, 1 (1969)
26. Szydlik, P.P., Green, A.E.S.: Independent-particle-model potentials for ions and neutral atoms with $Z \leq 18$. Phys. Rev. A **9**, 1885 (1974)
27. Garvey, R.H., Jackman, C.H., Green, A.E.S.: Independent-particle-model potentials for atoms and ions with $Z \leq 54$
28. Wilets, L., Henley, E.M., Kraft, M., MacKellar, A.D.: Classical many-body model for heavy-ion collisions incorporating the Pauli principle. Nucl. Phys. A **282**, 341 (1977)

29. Wilets, L., Kirschbaum, C.L.: Classical many-body model for atomic collisions incorporating the Heisenberg and Pauli principle. Phys. Rev. A **21**, 834 (1980)
30. Cohen, J.S.: Quasiclassical effective Hamiltonian structure of atoms with Z = 1 to 38. Phys. Rev. A **51**, 266 (1995)
31. Cohen, J.S.: Extension of quasiclassical effective Hamiltonian structure of atoms through Z = 94. Phys. Rev. A **57**, 4964 (1998)
32. Guzman, F., Errea, L.F., Pons, B.: Two active-electron classical trajectory Monte Carlo methods for ion-He collisions. Phys. Rev. A **80,** 042708 (2009)
33. Montemayor, V.J., Schiwietz, G.: Dynamical target screening for two-active-electron classical-trajectory Monte Carlo methods for H^+ + He collisions. Phys. Rev. A **40**, 6223 (1989)
34. Becker, L.R., MacKellar, A.D.: Theoretical initial l dependence of ion-Rydberg-atom collision cross sections. J. Phys. B **17**, 3923 (1984)
35. Rakovoc, M.J., Schultz, D.R., Stancil, P.C., Janev, R.K.: On quantum-classical correspondence in classical studies of atomic processes. J. Phys. A **34**, 4753 (2001)
36. Schultz, D.R., Stancil, P.C., Rakovoc, M.J.: Classical/quantal correspondence in state-selective charge transfer for partially stripped ion impact on atomic hydrogen: improvements to the CTMC model. J. Phys. B **34**, 2739 (2001)
37. Frémont, F.: Quasiclassical treatment of the Auger effect in slow ion-atom collisions. Phys. Rev. A **96**, 032712 (2017)

Chapter 3
Classical Treatment of $A^{q+} + H$ Collisions

Abstract One-electron collisions systems $A^{q+} + H$ collisions are studied within CTMC model and compared with available experiments and quantum calculations. A large part of the chapter is devoted to $H^+ + H$ collisions, which is a decisive test to validate classical calculations. Total cross sections, as well as partial and differential cross sections, are discussed. Then, $A^{q+} + H$ collisions are examined.

3.1 Interest and Motivation

Inelastic collisional processes involving atom or ion projectiles and atomic hydrogen H are of considerable interest, experimentally and theoretically. Since H atom constitute 95% of our universe and are present in many structures and experiments on earth, the understanding of processes and the determination of probabilities and total, partial, and differential cross sections is necessary. On a practical point of view, such reactions are relevant in tokamak fusion reactors, interstellar medium or solar wind.

Since absolute total cross sections are very difficult to obtain experimentally with uncertainties smaller than 10%, typically, theories have to be included and improved to compare with experimental results, and to predict detailed quantities such as partial or multi-differential cross sections.

Many theoretical approaches were developed using quantum mechanics and resolution of the Schrödinger equation. $A^{q+} + H$ collisions are of great interest because there is only one active target electron, so that electron correlation in target wave functions, that may obscure details of the collision dynamics, has not to be taken into account.

One problem is not the determination of the total cross sections. Measured total cross sections can be reproduced by theories with a high or satisfactory degree of accuracy for a broad range of kinematic conditions. Even if the Distorted-Wave Born approximation (DWBA) yields good agreement with experiment, even if models like the Continuum-Distorted-Wave Eikonal-Initial State (CDW-EIS) or Three-Body-Distorted Wave (3DW) approaches give a better agreement, it was recognized that total cross sections are not ideally suited to test theory? Indeed, in the integration of

© The Author(s), under exclusive license to Springer Nature Switzerland AG 2021

F. Frémont, *Classical Treatment of Collisions Between Ions and Atoms or Molecules*, Springer Series on Atomic, Optical, and Plasma Physics 118,
https://doi.org/10.1007/978-3-030-89428-3_3

differential cross sections over kinematical parameters, inaccuracies can be averaged out, leading to fortuitous agreement. In fact, it often appears that calculated doubly or triply differential cross sections are far to reproduce experiment, even for the simple collision system $H^+ + H$. Consequently, it is easily understandable that solving these problems using CTMC methods is a challenge to raise and an opportunity to seize. As mentioned in the introduction, even if quantum mechanics is the true method to apply, classical mechanics has the advantage that it can be used for any system at any projectile energies, except for $H^+ + H$ collisions at low energies.

In the following, the particular case of $H^+ + H$ collisions is studied. Then, extension to multiply charged ions on H is made.

3.2 $H^+ + H$ Collisions

The collision system $H^+ + H$ is expected to be the simplest one to study. Classically, this system was examined in detail, using several initial conditions and distributions. The first study was made by Abrines and Percival in 1966, and the most recent one was performed in 2010 for doubly differential cross sections.

3.2.1 Total Cross Sections

3.2.1.1 Experimental Cross Sections for SC and SI

Figures 3.1and 3.2 summarize the evolution of SC and SI total cross sections, respectively, obtained experimentally (full objects) (for more details, see [1] and references therein). In addition, an exponential function was used to fit the experiment for both

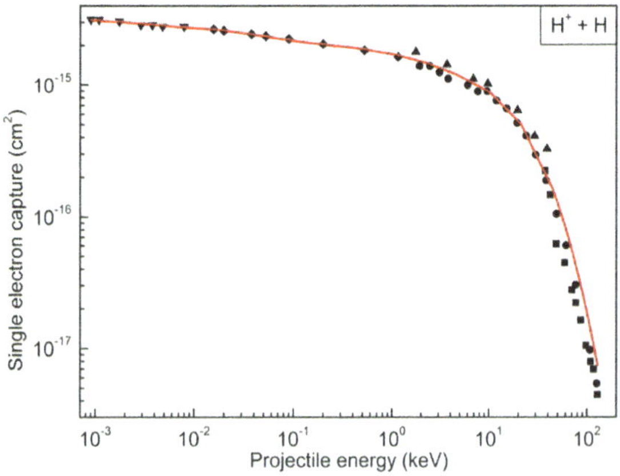

Fig. 3.1 Total experimental capture cross sections following $H^+ + H$ collisions ([1] and references therein) as a function of the projectile energy. The full red curve is a fit of the latter cross sections using exponential functions

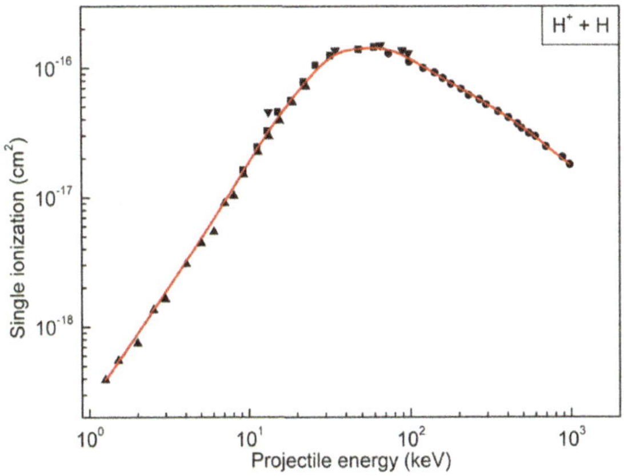

Fig. 3.2 Total experimental ionization cross sections following $H^+ + H$ collisions ([2, 3] and references therein) as a function of the projectile energy. The full red curve is a fit of the latter cross sections using exponential functions

cases (red full curve) and will be used in the next sections for comparisons with theory.

In the case of SC, cross sections are rather independent of the projectile energy below 1 keV and decrease rapidly at energies larger than 10 keV. For SI, cross sections reach a maximum at about 60 keV and decrease very rapidly around this value.

3.2.1.2 Comparison Between Experiment and Quantum Calculations

Figures 3.3 and 3.4 show comparison between experimental cross sections (full points derived from red curves of Figs. 3.1 and 3.2), for SC and SI, respectively. In

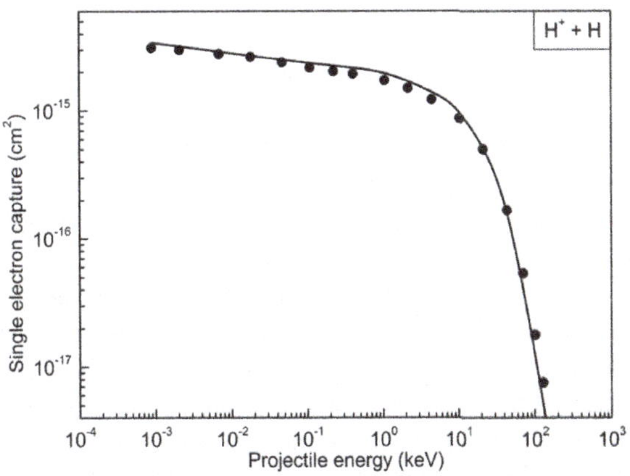

Fig. 3.3 SC total cross sections following $H^+ + H$ collisions as a function of the projectile energy. Experiment, derived from fit of experimental results (red curve in Fig. 3.1), full circles; coupled-channel approach [4], full curve

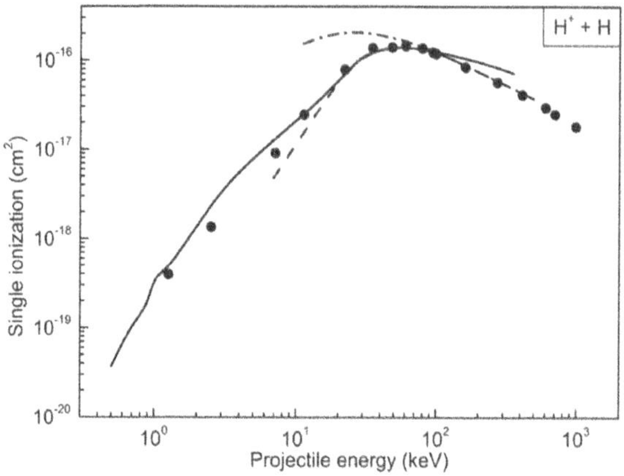

Fig. 3.4 SI total cross sections following $H^+ + H$ collisions as a function of the projectile energy. Experiment, derived from fit of experimental results (red curve in Fig. 3.2), full circles; theoretical calculations [2, 5], dashed, dotted dashed and full curve

the case of SC, the best agreement is observed for the model that used momentum-space integral-equation formalism [4] (full curve), while other theoretical results (not shown here) are only valid in certain regions of projectile energies.

For SI, no quantum theory is able to reproduce the experimental cross sections over the whole projectile energy domain. Born calculations [2] (dashed-dotted curve) is only valid at energies larger than 100 keV. CDW-EIS calculations [2] (dashed curve) reproduce the shape of SI cross sections but underestimate cross sections at energies smaller than 50 keV. In contrast, the two-center atom-state approximation [5] (full curve) model overestimate cross sections at energies larger than 50 keV.

3.2.1.3 Comparison Between Experiment and CTMC Calculations

Whatever the initial radial and momentum distributions used for the target electron, SI cross sections calculated by CTMC methods [6, 7] agree well with experiment (full circles) within a large range of projectile energies (Figs. 3.5 and 3.6 for SC and SI, respectively).

On the contrary, no classical method [6, 8] is able to reach experimental values at energies smaller than 20 keV (Fig. 3.5). According to some authors, this disagreement is due to the fact that $H^+ + H$ is a pure quantum system. Molecular aspects that occur at low energies are not taken into account; the tunnel effect is also missing in the calculations. Finally, the two protons are indiscernible, so that interference effects that are present during the collision are missing in CTMC calculations.

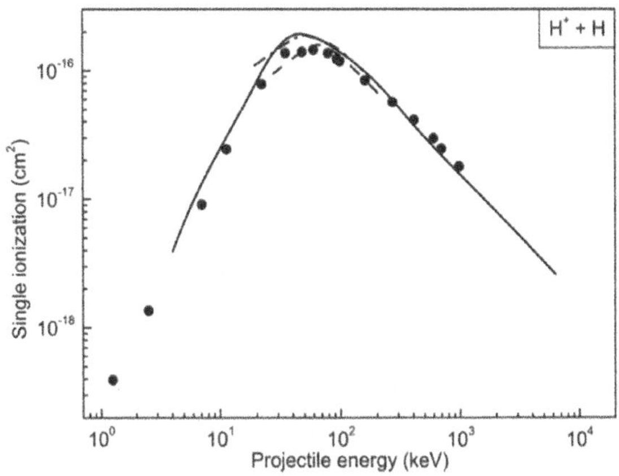

Fig. 3.5 SI total cross sections following $H^+ + H$ collisions as a function of the projectile energy. Experiment, derived from fit of experimental results (red curve in Fig. 3.2), full circles; CTMC calculations: microcanonical initial distribution [6], dotted-dashed curve; cut-off Wigner distribution [6], dashed curve; [7] full curve

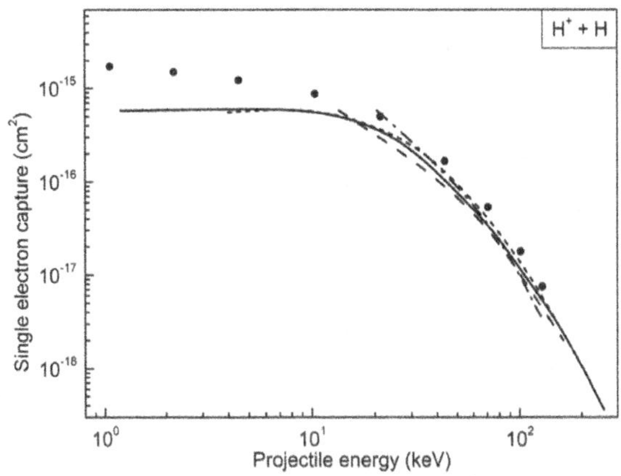

Fig. 3.6 SC total cross sections following $H^+ + H$ collisions as a function of the projectile energy. Experiment, derived from fit of experimental results (red curve in Fig. 3.1), full circles; CTMC calculations: [6], dotted-dashed curve; [6], dashed curve; [7], short dashed curve; [8], full curve

3.2.2 Partial Cross Sections

3.2.2.1 Electron Capture

As shown in the previous chapter, the target electron is captured not only onto the projectile fundamental state, but also on excited states. As mentioned before, using relation (2.51), partial cross sections σ_{SC}^{n} for SC on a state with principal quantum

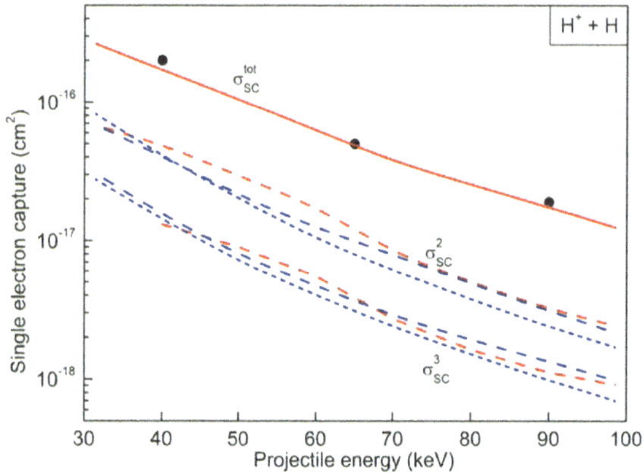

Fig. 3.7 Electron-capture partial cross sections for the H^+ + H system. The CTMC values are given by the full and dashed red curves for total and partial cross sections, respectively. Experiment (full points), Total cross sections [9]; quantum calculations (dashed and short dashed blue curves) [9]

number n is calculating by cutting the final energy distribution of the electron. An example is given in Fig. 3.7, in the projectile range from 30 to 100 keV Capture mainly occurs on $n = 1$, but also on excited states with $n > 1$. The agreement quantum (blues dashed curves) [9] and classical (red dashed curves) [9] calculations is rather good. For comparison are also represented total experimental cross sections (full circles) and CTMC results (full red curve) shown above (cf. Sect. 3.2.1).

The next step to check the validity of the CTMC model is calculations of the angular momentum l distributions. As an example, in H^+ + H collisions (Fig. 3.8), the electron can be captured on 2s or 2p levels ($l = 0$ or 1, respectively). Again, the agreement between CTMC, calculations (red curves) and experiment (full circles) and quantum calculations (blue curves) is fairly good. This allows us to use the classical model in order to analyze and understand more complex systems.

3.2.2.2 Target Excitation

Cross sections for target excitation in $H^+ + H$ collisions are necessary for calculating line profiles and intensities, in particular at intermediate and large projectile velocities, where all the inelastic processes compete.

Many experiments have been conducted to measure absolute excitation cross sections σ^{n}_{exc} for populating a specific n state, and σ^{nl}_{exc} for excitation on a specific l level for a fixed value of n (see [10] and references therein). Figure 3.9 shows recommended data (black full curves) for excitation to n = 2–4 (black curves). In parallel, excitation cross sections have been calculated using a variety of highly accurate quantum methods in a large range of projectile energies (see for example [11, 12]) and CTMC model [13]. Cross sections obtained by close-coupling two-center

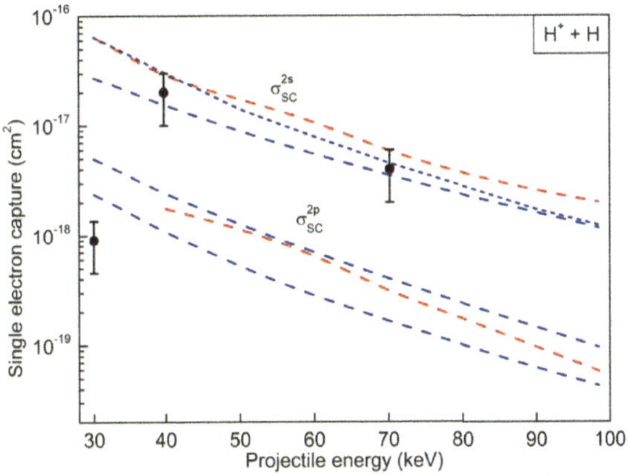

Fig. 3.8 Electron-capture partial cross sections for the $H^+ + H$ system into 2s and 2p states. The CTMC values are given by the full and dashed red curves for total and partial cross sections, respectively. Experiment (full points), quantum calculations (dashed and short dashed blue curves) [9]

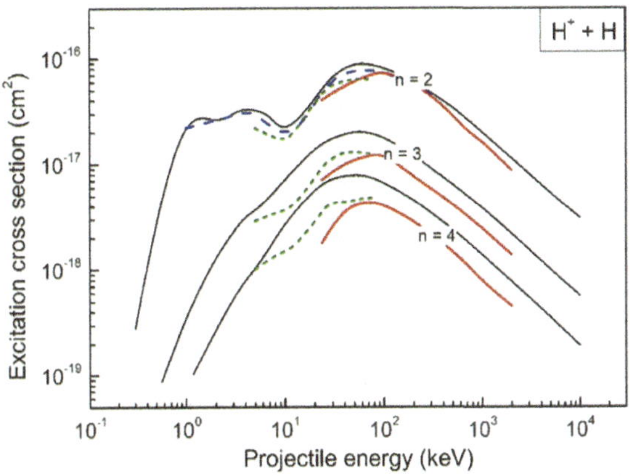

Fig. 3.9 Partial excitation cross sections following $H^+ + H$ collisions at projectile energies ranging from 1 keV up to 10^4 keV, into n = 2–4 states. Recommended data, black full curves [10]; quantum calculations [12] (blue dashed curve) and [11], green short-dashed curves; CTMC calculations, full red curves [13]

basis generator method [12] (blue dashed curve) are in excellent agreement with data, whereas those calculated by solving the Schrödinger equation with variable resolution and utilizing a hybrid spatial-Fourier grid (green short-dashed curves) deviate from data by a factor up to 2, especially for $n = 3$ and 4. CTMC calculations [13] (red full curves)), that are expected to be valid at high energies are in agreement with data for $n = 2$ but noticeably deviate for higher values of n.

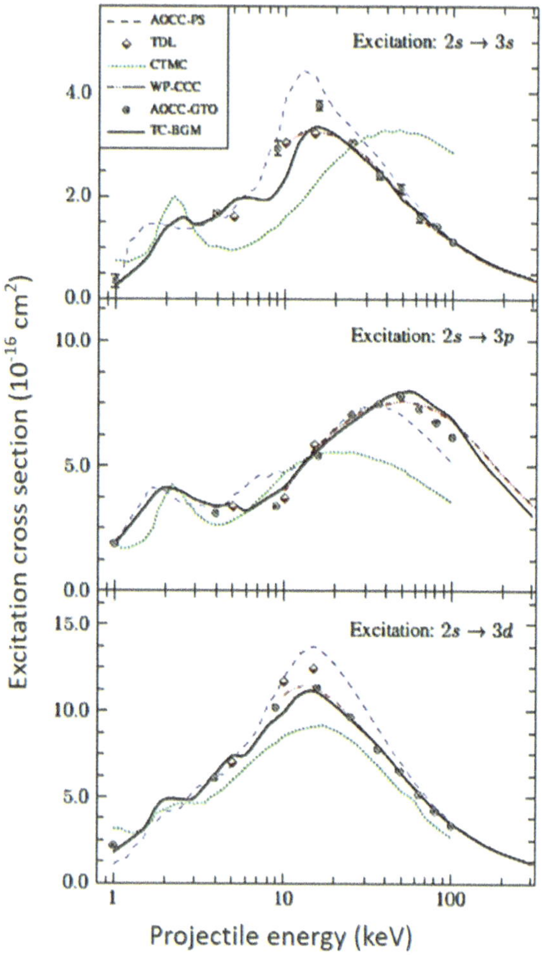

Fig. 3.10 Partial excitation cross sections σ_{exc}^{nl} following H^+ + H(2s) collisions at projectile energies ranging from 1 keV up to 10^4 keV, into n = 3*l* states [12]. CTMC calculations are represented by short-dashed green curves, whereas the other curves originate from quantum calculations. The calculations are compared with various experimental results (open diamonds and circles. For more details see [12] and references therein)

When entering into details, for example calculating σ_{exc}^{nl} (Fig. 3.10) from 2s state to 3*l* states, the disagreement is clearly visible for the CTMC model [12], whereas quantum calculations reproduce quite well the experiment.

3.2.3 Differential Cross Sections

Since many works have been devoted to calculating single and multiply differential cross sections, only some examples are given in this Section. Arbitrarily, differential cross section in impact parameter, electron ejection angle and energies will be discussed.

3.2.3.1 Impact Parameter Distributions

Impact parameter distributions, or differential cross sections in impact parameter b, $d\sigma/db = 2\pi b P(b)$, where $P(b)$ is the probability for a specific process, are the first step, in the impact parameter model, to deduce total and partial cross sections, by integration over b.

It is difficult to be confident in impact parameter distributions since b is not an observable quantity. Indeed, there is no unique correspondence between the projectile detection angle θ_d and b. More precisely, it is well known that a fixed value of θ_d corresponds to a wide distribution of b. In the case of $H^+ + H$ collisions, only calculations are available.

Figure 3.11 shows impact parameter distributions calculating by means of quantum mechanics [14] (black short-dashed curves) at projectile energies of 10 keV, 40 keV and 100 keV, and classical mechanic [8]. In CTMC calculations, micro-canonical distributions (red dashed curve) and phenomenological potentials (red full curve) were used. CTMC distributions have maxima that are located at impact parameter of the order of 2, 1.5 and 1 a.u., as the projectile energy decreases, whereas

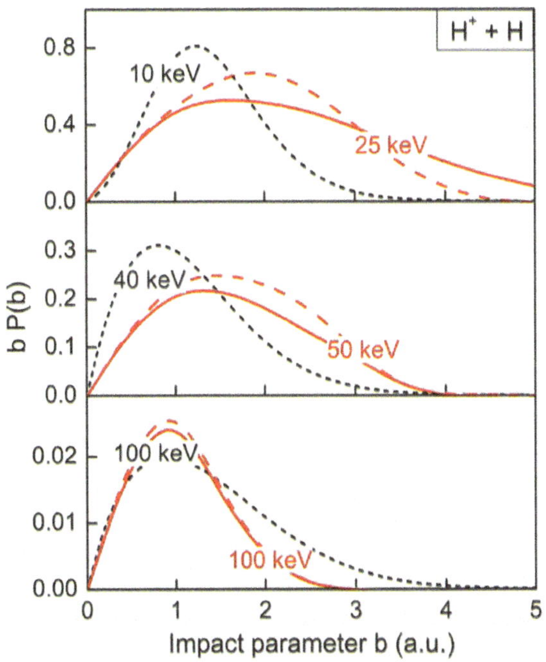

Fig. 3.11 Impact parameter distributions following electron capture in $H^+ + H$ collisions, for different projectile energies. Quantum calculations [14], black short-dashed curves for projectile energies of 10, 40 and 100 keV. CTMC calculations [8], red dashed and full curves, using micro-canonical distributions and phenomenological potential, respectively, for projectile energies of 25, 50 and 100 keV

quantum distributions present maxima around 1 a.u. In addition, classical distributions are much wider than those obtained with quantum mechanics, except at the highest energy. Finally, one concludes that the agreement between theories is more pronounced at the highest energies.

3.2.3.2 Scattering Angular Distributions

Quantum and classical methods have been applied to single differential cross sections (SDCS) $d\sigma/d\Omega_d$, with $d\Omega_d = 2\pi \sin\theta_d d\theta_d$, where θ_d is the scattering angle, or detection angle of the projectile, after single capture. It is noted that θ_d is related to b but, in contrast with b distributions, θ_d distributions can now be compared with experiment. As an example, Fig. 3.12 shows θ_d distributions following electron capture at a projectile energy of 125 keV [15, 16].

Experimental cross sections decrease from 10^{-11} cm^2/sr down to 3 10^{-14} cm^2/sr in the range 0–0.2 mrad. Quantum calculations (dashed, short dashed and dotted dashed curves) noticeably deviate from experiment, while CTMC calculations (red full curve) reproduce quite well the experiment, within the uncertainties, that are not shown in the Fig. 3.12.

The excellent agreement using CTMC acts as a decisive test to calculate other differential cross sections, such as angular distributions of ionized electrons, or electron energy distributions.

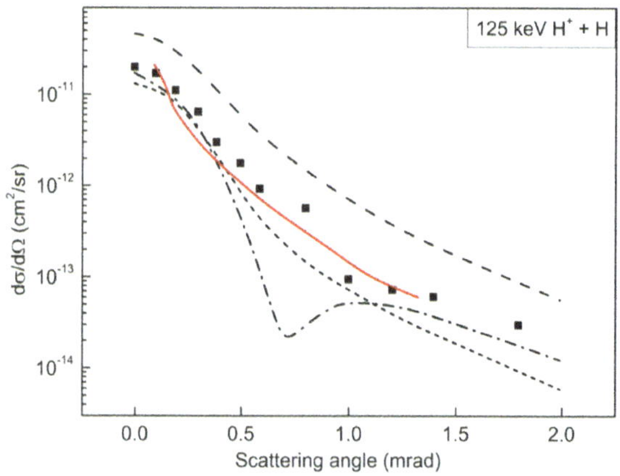

Fig. 3.12 SDCS for electron capture following electron capture in 125 keV H^+ + H collisions. Experiment, full squares; result of quantum calculations, dashed, short dashed and dotted dashed curves; CTMC calculations, red full curve [15, 16]

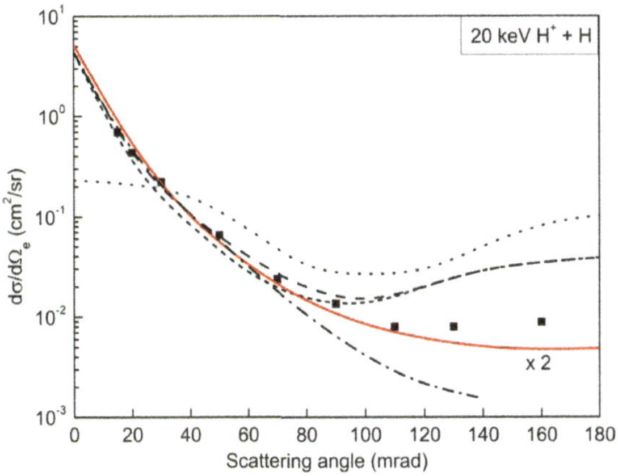

Fig. 3.13 Differential cross section $d\sigma_I/d\Omega_e$ following ionization in 20 keV $H^+ + H$ collisions. Experiment, full squares [17]; quantum calculations, black curves [17, 18]; red full curve, CTMC calculations [17]

3.2.3.3 Angular Distributions of Ionized Electrons

In Fig. 3.13 are plotted experimental differential cross sections (full squares) $d\sigma_I/d\Omega_e$, with $d\Omega_e = 2\pi \sin\theta_e d\theta_e$, where θ_e is the detection angle of the emitted target electron after ionization, for a projectile energy of 20 keV [17]. They are compared to a large variety of theoretical calculations, including quantum models (black curves) [17, 18] and CTMC model (red curve). No quantum theory is able to reproduce the experimental data, especially at large detection angles. Even the most sophisticated model which includes interferences (dashed curve) overestimates the experiment at angles larger than 90°. In contrast, whereas CTMC calculations do not reproduce the absolute values of the cross sections, it is clearly seen that, by multiplying the classical results by a factor of 2, the shape is close to that of experimental distribution. Note that, for higher projectile energies, CDW-EIS calculations agree very well with experiment, whereas CTMC calculation underestimate cross sections at large angles.

3.2.3.4 Electron Energy Distributions

A far more stringent test for quantum and classical theories is to calculate multiply differential cross sections (MDCS), in particular doubly or triply differential cross sections (DDCS or TDCS) since they contain much more information on the processes themselves. Such DDCS were performed experimentally, by measuring $d^2\sigma_I/d\Omega_e dE_e$, where E_e is the energy of the target electron after ionization. Figure 3.14 presents an example for DDCS at a projectile energy of 114 keV, for which calculations are expected to be valid, and at emission angles of 20°, 50°, 90° and 130°. CTMC calculations (red full curve) and CDW-EIS calculations (dashed curves) reproduce quite well the experiment, whereas serious discrepancies are visible for first Born approximations (short dashed curves).

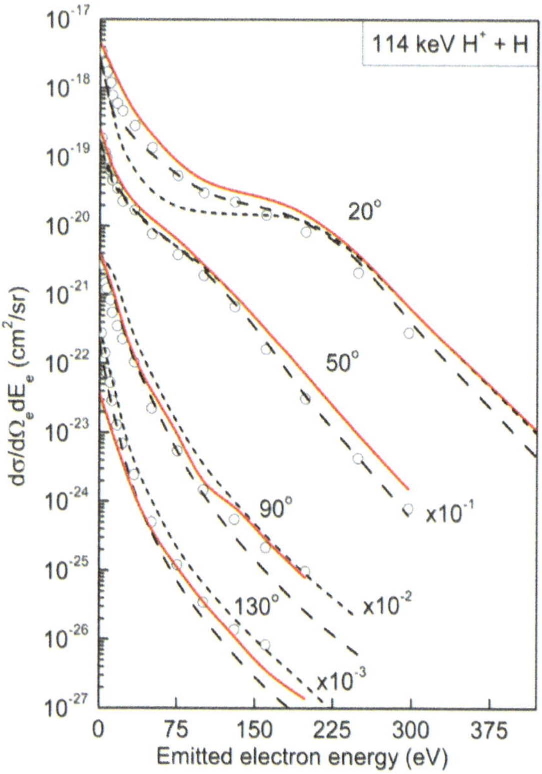

Fig. 3.14 DDCS $d^2\sigma_I/d\Omega_e dE_e$ following ionization in 114 keV H^+ + H collisions. Experiment, empty circles; Quantum calculations, dashed curves (CDW-EIS) and short dashed curves (Fiste Born); CTMC calculations, full red curve

3.3 A^{q+} ($q > 1$) + H Collisions

In view of the ability for CTMC calculations to reproduce quite well experimental data in the case of H^+ + H collisions, and since CTMC method is easier and faster to implement than quantum calculations, many classical calculations were performed in a wide range of projectile charges and velocities, to study elastic and inelastic processes in heavy charge d particles A^{q+} + H collisions.

3.3.1 Total Cross Sections

The first systematic classical study was done in 1977 for C, N, O and Ar projectiles [19], but only a few experimental results were available to compare with CTMC calculations. In 1981 [9], *nl* distributions following electron capture in A^{q+} + H collisions were determined for q varying from 1 to 20, but no comparison with experiment could be done.

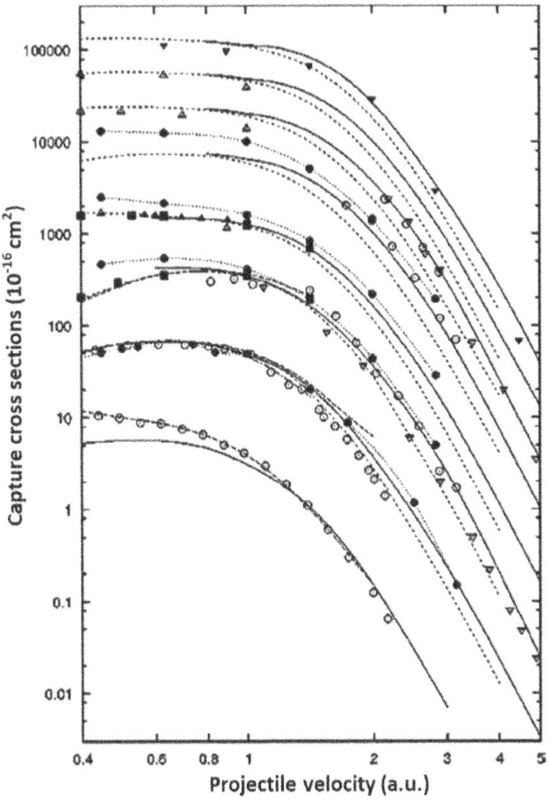

Fig. 3.15 Total capture cross sections following A^{q+} + H collisions, as a function of projectile velocities. CTMC calculations (full curves) are compared with various experiments and quantum calculations [20]

In 1999, a complete study of electron capture and ionization in A^{q+} + H collisions was performed using CTMC method in order to deduce total cross sections [20]. The results are presented in Figs. 3.15 and 3.16 for capture and ionization, respectively. To improve calculations, discrete microcanonical functions rather than continuous functions were used, with the condition on the average energy of the target electron $< E > = -0.5$ a.u. Comparison with various experiments show an improving accord obtained as q increases. This improvement is also apparent with different theoretical approaches. Note that, for electron capture, CTMC cross sections are systematically higher than quantum cross sections for projectile velocities larger than 1 a.u.

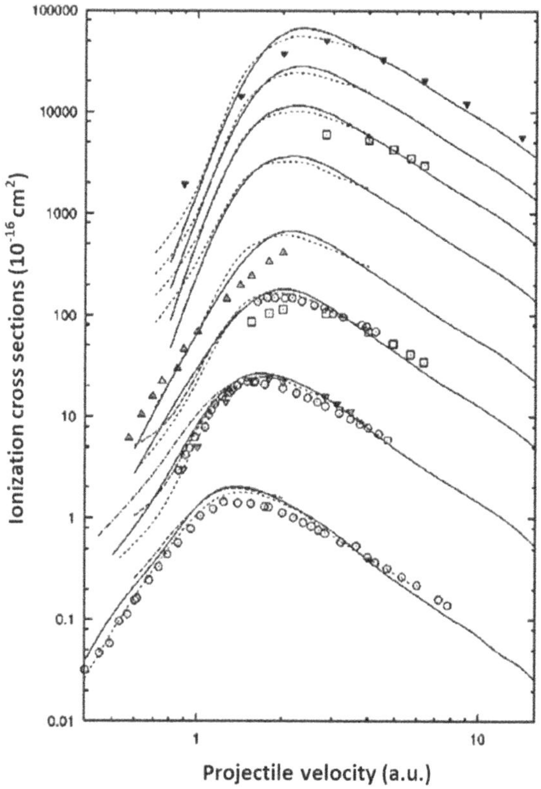

Fig. 3.16 Total ionization cross sections following A^{q+} + H collisions, as a function of projectile velocities. CTMC calculations (full curves) are compared with various experiments and quantum calculations [20]

3.3.2 *Differential Cross Sections*

3.3.2.1 Impact Parameter Distributions

In Fig. 3.17 are plotted differential cross sections *bP(b)* as a function of *b*, for Ne^{10+} + H collisions [21], at projectiles velocities of 0.4 a.u. (top of Fig. 3.17), 1 a.u. (middle of Fig. 3.17) and 2 a.u. (bottom of Fig. 3.17).

Full and dashed curves represent CTMC and OEDM quantum calculations, respectively. At the largest velocity, the agreement between both methods is very good. At lower velocities, the amplitudes of the quantum distributions are higher than the classical one, whereas widths are found to be smaller. At 0.4 a.u., oscillations appear in the quantum calculations that are not present in CTMC calculations. When integrating *bP(b)* over *b*, classical and quantum total cross sections are very similar, despite differences observed in *b* distributions. This agreement results of a cancellation of errors in the amplitudes and widths.

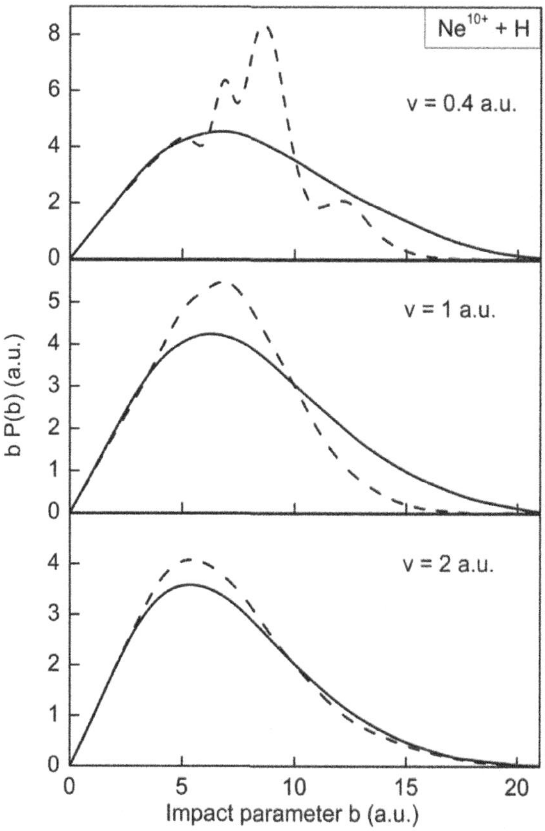

Fig. 3.17 Impact parameter distributions following Ne^{10+} + H collisions, at projectiles velocities of 0.4 a.u. (top of Fig. 3.17), 1 a.u. (middle of Fig. 3.17) and 2 a.u. (bottom of Fig. 3.17). Quantum OEDM calculations, dashed curves; CTMC calculations, full curves

3.3.2.2 Doubly Differential Cross Sections

In this section, we discuss a particular and fundamental point that has been analyzed for many years since 2001. Due to the development of experimental techniques that permit the coincidence of many particles, attention has been focused on processes responsible for single ionization of He and H following the impact of highly charged ions (especially C^{6+} and Au^{53+}) at different regimes. Since serious discrepancies between experiment and quantum and classical model arose in the case of Au^{53+} + He collisions, the mechanisms of ionization have been studied in details for H target, using quantum CDW and CTMC calculations [22]. The choice of H was motivated by two difficulties to avoid: Firstly, the choice of three-body model to describe a pure four-body problem is decisive in the comparison with experiment. Secondly, complications due to double ionization (DI) have been addressed.

In this work [21] TDCS $d^3\sigma_I/dQ_\perp d\Omega_e dE_e$ and DDCS $d^2\sigma_I/dQ_\perp dE_e$ have been investigated by means of CDW and CTMC calculations. The quantity $Q_\perp$ is the perpendicular component of the momentum transferred $\vec{Q}$ to the target by the projectile, which is defined by $\vec{Q} = \vec{p_o} - \vec{p_f}$, where $\vec{p_o}$ and $\vec{p_f}$ are the initial and final projectile momenta (Fig. 3.18).

Figure 3.19 shows $Q_\perp$ distributions in the case of C^{6+} projectiles impacting on H, for emitted electron energies of 10, 50 and 90 eV. Since H ionization mainly occurs at impact parameters b of the order of a few a.u., Low electron energies correspond to large impact parameters whereas, on the contrary, high energies are related to small impact parameters. At $E_e = 10$ eV, $Q_\perp$ distributions monotonically decrease

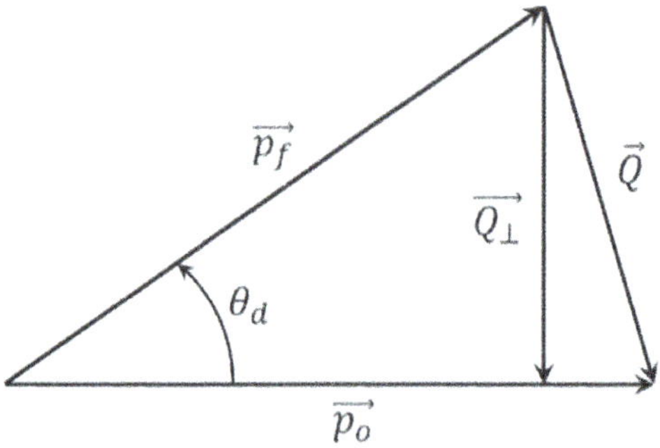

Fig. 3.18 Schematic representation of transferred momentum $\vec{Q} = \vec{p_o} - \vec{p_f}$, where $\vec{p_o}$ and $\vec{p_f}$ are the initial and final projectile momenta. The vector $\vec{Q_\perp}$ is the perpendicular component of $\vec{Q}$ relative to the quantization axis. The quantity θ_d is the scattering angle

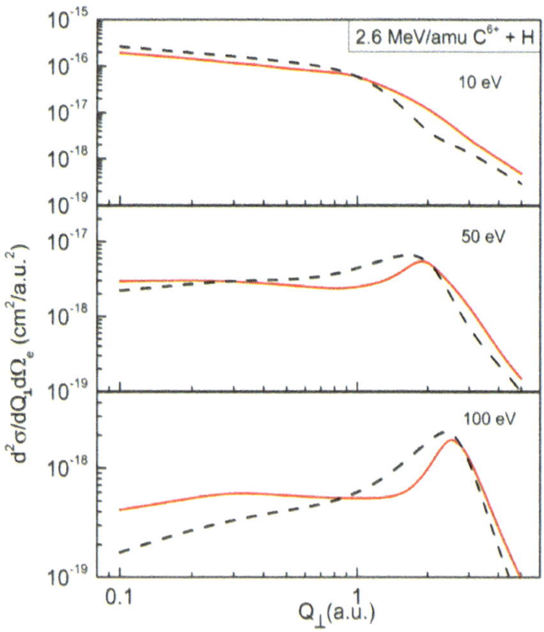

Fig. 3.19 $Q_\perp$ distributions following 2.6 meV/amu C^{6+} + H collisions, for emitted electron energies of 10, 50 and 90 eV. CDW calculations, dashed curves; CTMC calculations, full red curves [22]

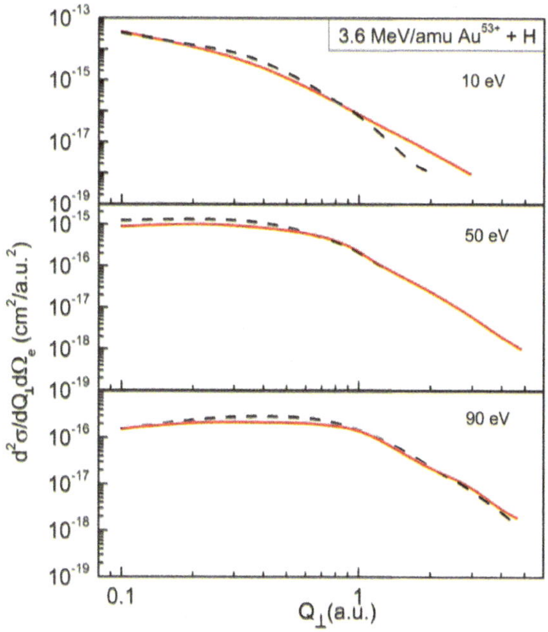

Fig. 3.20 $Q_\perp$ distributions following 3.6 meV/amu Au^{53+} + H collisions, for emitted electron energies of 10, 50 and 90 eV. CDW calculations, dashed curves; CTMC calculations, full red curves [22]

with increasing $Q_\perp$. At larger values of E_e, a peak centered at $Q_\perp$ ~ 2–3 a.u. is observed, due to binary collisions, i.e. direct collisions between the projectile and the target electron. CDW and CTMC are found to be in good agreement, except at 10 eV for large values of $Q_\perp$, showing that classical methods are able to give detailed information on the collision.

In Fig. 3.20 are shown DDCS for ionization of H by 3.6 meV/amu by Au^{53+} ions. Firstly, we have to mention the remarkable agreement between quantum (dashed curves) and classical (full curves) calculations. The shapes of the distributions differ from those observed for C^{6+} ions obtained at the same emitted electron energies. A dominance of low momentum transfer is observed whatever the value of E_e. In contrast with C^{6+} impact, no binary peak is visible. The main reason is that ionization takes place mainly at large impact parameters, so that binary collisions, even if they contribute to cross sections, are less probable than three-body (or soft) collisions. It is evident that experiments are missing to obtain valuable conclusions. However, we will see in the next chapter that experimental results are not necessarily reliable to definitively conclude on the mechanisms responsible for ionization, and that CTMC calculations are an excellent tool to validate unexpected hypothesis.

3.4 The Role of Interferences in Collisions

3.4.1 Description of Interferences in Classical Mechanics

The existence and the role of interferences in collisions no longer needs to be demonstrated (see [23] and references therein). The principle of interferences and conditions to obtain and visualize them have been summarized by R. Feynman at the beginning of the 1960's [24]. Suppose to different ways for a particle that will be detected at a certain position with a certain energy. If we do not try to find out which way the particle took before it is detected, interferences are created and can be detected and observed. On the contrary, if we try to follow the trajectory of the particle, the interference pattern will be lost.

In practice, the interferences manifest themselves in oscillations, for example in energy distributions, impact parameter distributions, angular distributions, or in the dependence of cross sections with projectile velocity. In principle, classical mechanics is in essence impossible to reveal interferences, since classical matter is only considered as particles, and has no wave component. Nevertheless, it has been recently shown that wave behaviour can be revealed, by looking for the phase difference induced by the delay of one particle over another one, as shown in Fig. 3.21.

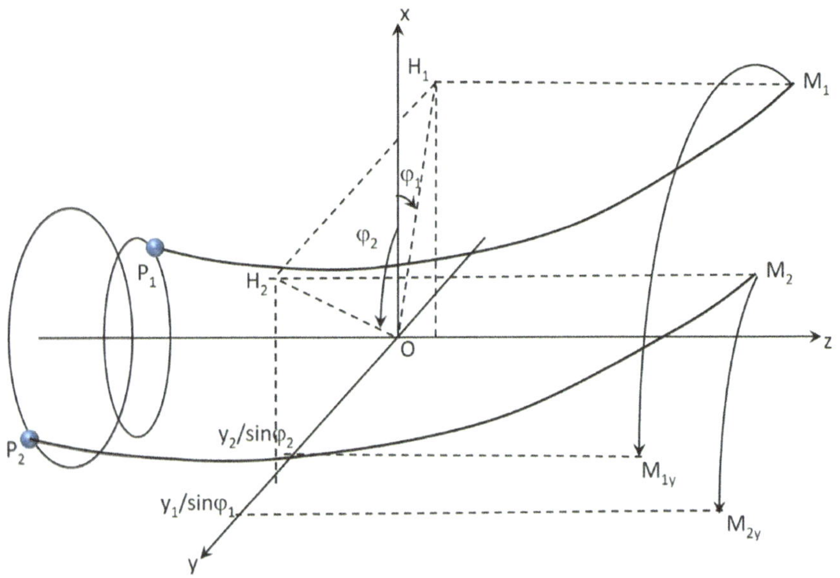

Fig. 3.21 3D schematic view of two particle (P_1 and P_2) trajectories in the field of a target, located at the origin O of the reference frame. At final time t_{max}, the particles are on P_1 and P_2. Their projection on (xOy) plane after rotation are M_{1y} and M_{2y}, respectively

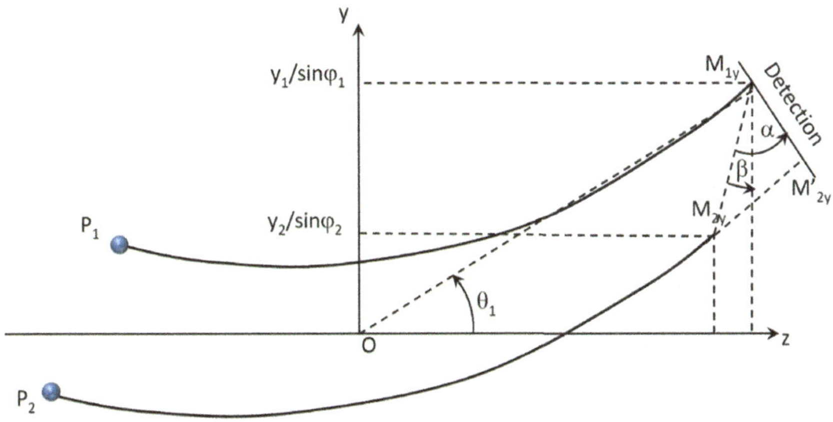

Fig. 3.22 2D trajectory schematic view of the two particles. The delay is characterized by the distance$M'_{2y}M_{2y}$

Suppose two particles P_1 and P_2 having independent trajectories, scattering on a target initially located on O. At $t = t_{max}$, the particles reach the detector at positions M_1 and M_2, respectively. Since there is a symmetry around axis *(Oz)*, the points M_i are artificially rotated around *(Oz)* in order to be projected on the *(yOz)* plane, becoming $M_{1\mathrm{y}}$ and $M_{2\mathrm{y}}$. Projections of M_1 and M_2,on *(Oz)* are unchanged, whereas projections on *(yOz)* are $y_1/\sin\varphi_1$ and $y_2/\sin\varphi_2$, respectively.

Let us resume the trajectories on *(yOz)* plane (Fig. 3.22). Suppose that the particle P_1 reaches the detector before P_2, and that their velocities at detection time are collinear, making angles θ_1 and θ_2 that are equal within $\Delta\theta$. The delay between both particles is $\delta_{12} = M'_{2y}M_{2y}$. This delay is easily calculated using simple geometry, by introducing the α and β angles.

With the condition that the probability for one particle to reach the detector after the collision is equal to unity, the amplitude associated with the particle is given by [23]

$$A_i = e^{-iE_i\left(t_{max}-\frac{\delta_{1i}}{v_i}\right)} \tag{3.1}$$

where E_i and v_i are the energy and the velocity of particle *i*, respectively.

The total amplitude at a fixed angle is thus the sum of the individual amplitudes A_i. Finally, the intensity is determined by the square modulus of the total amplitude.

To illustrate the mechanisms that are responsible for oscillations, two specific collisions are discussed below.

3.4.2 Interferences in $H^+ + H$ Collisions

At very low projectile velocities, oscillatory structures have been evidenced in H^+ + H collisions in the impact velocity distribution of capture probability [25]. Similar oscillations also appear in b distributions, as well as in the projectile scattering angle distributions. Whereas quantum mechanics interprets oscillations in terms of interferences between the main capture and the elastic scattering channels, classical mechanics interprets them in terms of swaps of the electron between the projectile and the nuclear center.

To check the different interpretations, $H^+ + H$ collisions have been revisited at an impact energy of 250 eV. Attention has been focused on the angular distribution of H atoms after electron capture [26].

In the case of a two-body problem, the differential capture cross section $d\sigma_C/d\Omega_d$ would write

$$\frac{d\sigma_C}{d\Omega_d} = \frac{b}{\sin\theta_d}\left|\frac{db}{d\theta_d}\right| \tag{3.2}$$

where θ_d is the detection angle and $\Omega_d = 2\pi\sin\theta_d d\theta_d$.

In the present case, there is no biunivocal relation between b and θ_d, since this is a true three-body problem, so that $db/d\theta_d$ is not defined. Hence b has to be replaced, at a given angle θ_d, by an impact parameter distribution $f(b) = N_c(b)/N_{tot}(b)$, where $N_c(b)$ and $N_{tot}(b)$ are the number of captured electrons and the total number of projectiles involved in the collision, respectively. Thus, relation (3.2) becomes

$$\frac{d\sigma_C}{d\Omega_d} = \frac{1}{\sin\theta_d d\theta_d}\int_{b_{min}}^{b_{max}} bf(b)db \tag{3.3}$$

Figure 3.23 shows the classical results for $d\sigma_C/d\Omega_d$ for scattering angles varying from 0.05° up to 7°. The black dashed curve is the result of the calculation if relation (3.2) is used. Even within this approximation, series of maxima and minima are visible, showing evidence for the classical interpretation of oscillations. The results are nearly unchanged if relation (3.3) is used (blue curve and open squares in Fig. 3.23a). One of the differences is that maxima (indicated by arrows) and minima are more visible than previously and shifted when compared with 2-body calculations. When interferences between capture pathways are introduced by applying the method escribed in Sect. 3.4.1 (red curve and open squares in Fig. 3.23b), the oscillations are also observed, with a larger amplitude than that observed in Fig. 3.23b.

The present calculations were found to be in agreement with previous experiments and quantum calculations. They show that both interpretations (number of swaps the electron experiences before capture, and interferences of two different collisional paths) are reasonable, and not opposite.

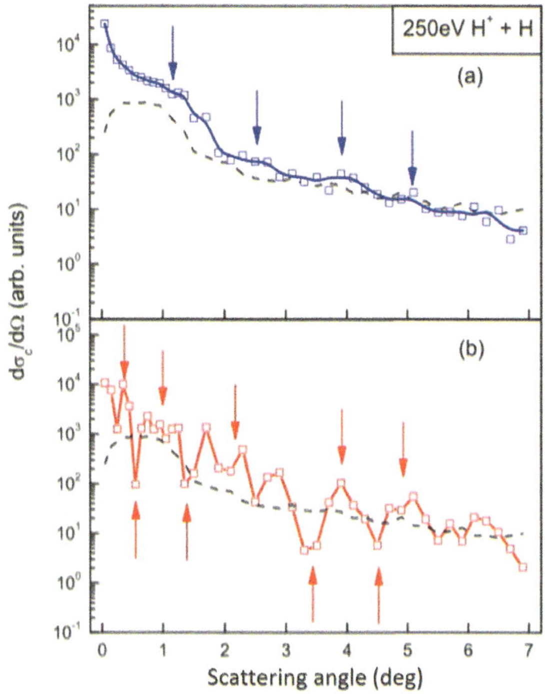

Fig. 3.23 Calculated differential capture cross sections $d\sigma_C/d\Omega_d$ as a function of the scattering angle. Dashed black curve, classical calculations using relation (3.1); blue curve and open squares, calculations using relation (3.2); red curve and open squares, calculations using relation (3.2) including interferences

3.4.3 Interferences in He^{2+} + H Collisions

The question of oscillations has not been only asked for resonant systems, such as $H^+ + H$ system, but also in the case of total cross sections following inelastic collisions involving highly charged ions. Even if the classical interpretation was well known to exist, these oscillations were often found to vanish after integration of differential cross sections. This is the case for example after integration over the impact parameter. To avoid the problem, the choice was focused on reaction channels which become predominantly populated at small impact parameters. The collision system $He^{2+} + H$ is a good candidate [27], since the maximum value for b is smaller than 2 a.u.

In Fig. 3.24 are displayed excitation cross sections σ_{exc} into $n = 2$ as a function of the impact energy. Quantum calculations (dashed and shot-dashed curves) are compared with CTMC calculations (full red curve) and experiment (full circles) as a function of the projectile energy in the range from 2 keV up to 600 keV. Oscillations are clearly visible. Two maxima appear at ~20 keV and 60 keV in both experimental and theoretical results. At energies smaller than 20 keV, CTMC calculations noticeably deviate from quantum calculations. However, no experimental results are available in this range, so that definitive conclusion is not possible.

Fig. 3.24 Excitation cross sections following He^+ + H collisions as a function of the projectile energy [27]. Experiment, full circles; dashed and short-dashed curves, results of quantum calculations; red full curve, CTMC calculations

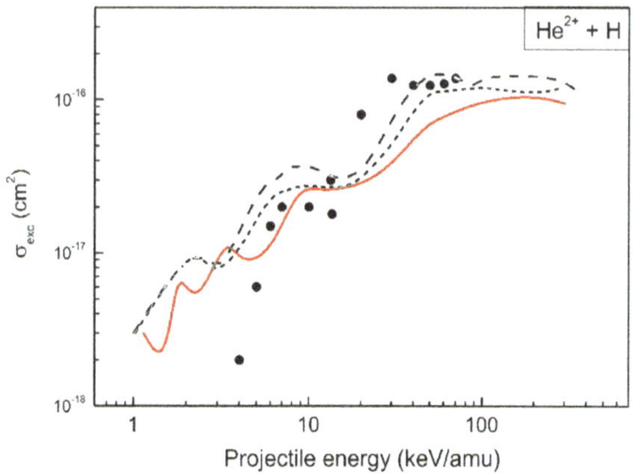

As in the case of the resonant system H^+ + H, these oscillations were found to be associated with the number of swaps that the electron undergoes during the collision. Again, even if interpretation differs whether classical or quantum mechanics are used, it is clearly demonstrated that CTMC calculations are a useful tools to describe in details collision processes for one electron systems.

References

1. Kadyrov, A.S., Bray, I., Stelbovics, A.T.: On-shell coupled-channel approach to proton-hydrogen collisions without partial-wave expansion. Phys. Rev. A **73** 012710 (2006)
2. Sahoo, S., Roy, K., Sil, N.C., Mukherjee, S.C.: Ionization of atomic hydrogen by slow protons. Phys. Rev. A **59**, 275 (1996)
3. Abdurakhmanov, I.B., Kadyrov, A.S., Avazbaev, S.K., Bray, I.: Solution of the proton-hydrogen scattering problem using a quantum mechanical two-center convergent close-coupling method. J. Phys. B **49** 115203 (2016)
4. Kadirov, A.S., Bray, I., Stelbovics, A.T.: On-shell coupled-channel approach to proton-hydrogen collisions without partial-wave expansion. Phys. Rev. A **73** 012710 (2006)
5. Toshima, N.: Convergence and completeness of the pseudostate expansion for proton-hydrogen collision in two-center close-coupling calculations. Phys. Rev. A **59**, 1981 (1999)
6. Eichenauer, D., Grün, N., Scheid, W.: Classical trajectory calculations for H^+ + H collisions with the Wigner functions as initial phase distribution. J. Phys. B **14**, 3929 (1981)
7. Illescas, C., Riera, A.: Classical study of single-electron capture and ionization processes in A^{q+}+(H, H2) collisions. Phys. Rev. A **60**, 4546 (1999)
8. Cohen, J.S.: Classical phase-space distributions for atomic hydrogen collisions. J. Phys. B **18**, 1759 (1985)
9. Olson, R.E.: n, l distributions in Aq^+ + H electron capture collisions. Phys. Rev. A **24**, 1726 (1981)
10. Janev, R.K., Reiter, D., Samm, U.: Collisions processes in low-temperatre hydrogen plasmas, Berichte des Forschungzentrums Jülich, vol. 4105 (ed. Forschungzentrums, Zentralbibliothek) (2003)
11. Tseliakhovich, D., Hirata, C.M., Heng, K.: Excitation and charge transfer in H-H+ collisions at 5–80 keV and application to astrophysical shocks. Mon. Not. R. Astron. Soc. **422**, 2357 (2012)

12. Leung, A.C.K., Kirchner, T.: Proton impact on ground and excited states of atomic hydrogen. Eur. Phys. J. D **73**, 246 (2019)
13. Reinhold, C.O., Olson, R.E., Fritsch, W.: Excitation of atomic hydrogen by fully stripped ions. Phys. Rev. A **41**, 4837 (1990)
14. Kolakowska, A., Pindzola, M.S., Robichaux, F., Schultz, D.R., Wells, J.C.: Excitation and charge transfer in proton-hydrogen collisions. Phys. Rev. A **58**, 2872 (1998)
15. Toshima, N., Ishihara, T., Ohsaki, A., Watanabe, T.: Impact-parameter treatment of classical trajectory Monte Carlo calculations for ion-atom collisions. Phys. Rev. A **40**, 2192 (1989)
16. Alt, E.O., Kadyrov, A.S., Mukhamedzanov, A.M.: Three-body approach to proton-hydrogen charge exchange and elastic scattering. Phys. Rev. A **60**, 314 (1999)
17. Kirby, G.W., Gealy, M.W., Hsu, Y.-Y., Rudd, M.E.: Energy and angular distributions of electrons from impact on atomic and molecular hydrogen. II. 20–114 keV H^+ + H. Phys. Rev. A **51** 2256 (1995)
18. Reading, J.F., Fu, J., Fitzpatrick, M.J.: Finite-Hilbert-basis-set calculations for the angular distribution of ionized electrons produced in p + H impact at 20 keV. Phys. Rev. A **70** 032718 (2014)
19. Olson, R.E., Salop, A.: Charge-transfer and impact-ionization cross sections for fully and partially stripped positive ions colliding with atomic hydrogen. Phys. Rev. A **16**, 531 (1977)
20. Illescas, C., Riera, A.: Classical study of single-electron capture and ionization processes in A^{q+} +(H, H_2) collisions. Phys. Rev. A **60**, 4546 (1999)
21. Errea, L.F., Illescas, C., Mendez, L., Pons, B., Riera, R., Suarez J.: Accuracy of the classical trajectory Monte-Carlo method for electron capture in $C^{6+}$and C^{6+}+ H(1s) collisions. Phys. Rev. A **70** 052713 (2004)
22. Fiol, J., Olson, R.E.: Three-body dynamics in hydrogen ionization by fast highly charged particles. J. Phys. B **35**, 1759 (2002)
23. Frémont, F.: Young-type interferences with electrons: Basics and theoretical challenges in molecular collision systems, Springer series on at., opt. and plasma phys. (Springer-Verlag Berlin Heidelberg) **77** (2014)
24. Feynman, R., Leighton, R., Sands, M.: The Feynman lectures on physics. Addison-Westley, Redwood city (1964)
25. Houver, J.C., Fayeton, J., Barat, M.: Elastic and inelastic differential measurements for H^+ on H collisions in the 250 eV–2000 eV energy range. J. Phys. B **7**, 1358 (1974)
26. Frémont, F.: Semiclassical analysis of angular differential cross sections for single-electron capture in 250-eV H^+ + H collisions. Phys. Rev. A **91** 052709 (2015)
27. Schultz, D.R., Reinhold, C.O., Krstic, P.S.: Analysis of unexplained oscillations in intermediate-energy ion-atom collisions. Phys. Rev. Lett. **78**, 2720 (1997)

Chapter 4
Classical Treatment of A^{q+} + He Collisions

Abstract Since it is a pure five-body problem, the challenge of the study of A^{q+} + He collisions consists of taking into account electron correlation during the collision. Because quantum mechanics is complicated to develop for such problems, the classical mechanics is expected to be a good alternative.

4.1 Interest and Motivation

The quantum or classical treatment of processes occurring in A^{q+} + He collisions is a greater challenge than that in A^{q+} + H collisions, for the main reason that it is a four-body problem, involving two nuclei and the two He electrons, if we assume that the projectile electrons do not participate to the collision. When quantum mechanics is used, the electron–electron interaction, also called electron correlation, is difficult to take into account, and only a few works were devoted to the analysis of inelastic processes including this interaction. One of the problems is, even if the potential associated with electron correlation can be considered as a perturbation, it may lead to large cross sections. When CTMC method is used, electron correlation is very difficult to introduce, since its inclusion leads to an instability of *He*, except if phenomenological potentials are added in the Hamiltonian (cf. Chap 2, Sect. 2.4.3).

For a long time, independent electron models were developed, neglecting the electron–electron interaction. These models were very useful and have shown their ability to account for observed phenomena. Nevertheless, the agreement was found to be accidental, either at low energy where electron capture is dominant, and at high energy in the case of ionization. In fact, it has been shown that the domain of projectile energies in which the independent electron models well describe dross sections is ultimately small compared to the energy domain studied today. Even if electron correlation is not necessary to determine total cross sections with a good precision, this interaction is required to calculate partial or differential cross sections.

The goal here is to show the power of CTMC calculations to describe pure four-body collisions.

© The Author(s), under exclusive license to Springer Nature Switzerland AG 2021
F. Frémont, *Classical Treatment of Collisions Between Ions and Atoms or Molecules*,
Springer Series on Atomic, Optical, and Plasma Physics 118,
https://doi.org/10.1007/978-3-030-89428-3_4

4.2 Processes in A^{q+} + He Collisions

As mentioned in Chap. 2, the number of processes that occur during the collision is much larger than that for $A^{q+} + H$ collisions. Below, we summarize all possible processes. Since we are only interested by the target electrons, we suppose a fully stripped ion A^{Z+} as a projectile.

Elastic scattering (ES) (4.1) is the dominant process, giving rise to approximately 90% of the processes. Single excitation (SE) and single ionization (SI) are the main processes at high projectile energies. Double excitation (DE) and double ionization (DI) are also likely to occur at high energies, but with a probability of a few per thousand of SI and SE, depending on the charge of the projectile. Transfer ionization (TI) and transfer excitation (TE) are dominating the intermediate projectile energy range. Finally, at low projectile energies, single capture (SC) and double capture are the main processes. Double capture process can be separated in non-autoionizing double capture (NADC) and autoionizing double capture (ADC). Note that all the processes described below can be separated, except ADC and TE (for more details, see Chap. 2).

$$A^{Z+} + He \rightarrow A^{Z+} + He \quad \text{ES} \tag{4.1a}$$

$$A^{Z+} + He \rightarrow A^{Z+} + He^{*} \quad \text{SE} \tag{4.1b}$$

$$A^{Z+} + He \rightarrow A^{Z+} + He^{**} \quad \text{DE} \tag{4.1c}$$

$$A^{Z+} + He \rightarrow A^{Z+} + He^{+} + e^{-} \quad \text{SI} \tag{4.1d}$$

$$A^{Z+} + He \rightarrow A^{Z+} + He^{2+} + 2e^{-} \quad \text{DI} \tag{4.1e}$$

$$A^{Z+} + He \rightarrow A^{(Z-1)+} + He^{+} \quad \text{SC} \tag{4.1f}$$

$$A^{Z+} + He \rightarrow A^{(Z-2)+} + He^{2+} \quad \text{NADC} \tag{4.1g}$$

$$A^{Z+} + He \rightarrow A^{(Z-2)+} + He^{2+} \quad \text{TDC or TI} \tag{4.1h}$$

$$A^{Z+} + He \rightarrow A^{(Z-1)+} + He^{+*} \quad \text{TE} \tag{4.1i}$$

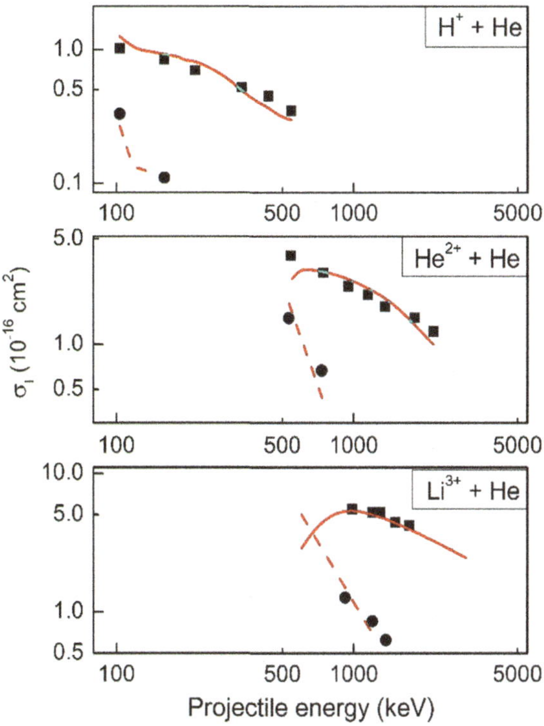

Fig. 4.1 Experimental and theoretical single ionization and single capture cross sections for A^{Z+} + He collisions (Z = 1, 2, 3) as a function of the projectile energy [1]. Experiment: SI, full squares; SC, full circles. Theory: SI, full red curves; SC, dashed red curves

4.3 Total and Partial Cross Sections

4.3.1 Total Cross Sections

The simplest way to describe CTMC calculations for A^{q+} + He collisions is to proceed chronologically. First, we focus on calculations that consider the target electrons as independent particles. The fist systematic study of electron loss[1] cross sections, including SI and SC, has been performed in 1978 [1], at projectile energies varying from 100 keV/amu up to 600 keV/amu, for projectiles with $Z = 1$–8. The extension of one-electron target to multielectron targets was made by determining the effective charge Z_{eff} of the target nucleus as seen by the active nucleus. The Hartree–Fock value of 1.69 was used in the calculations.

Since only a few experimental data exist for $Z > 3$, we focus on $Z = 1$, 2 and 3 (Fig. 4.1). Experimental cross sections are depicted by full squares (SI) and full circles (SC), whereas full and dashed red curves are CTMC results for SI and DI, respectively. The overall agreement between experiment and theory is clearly visible. Nevertheless, at the lowest energies, theoretical SI cross sections are seen to be

[1] In the following, electron loss will refer to ionization and electron capture.

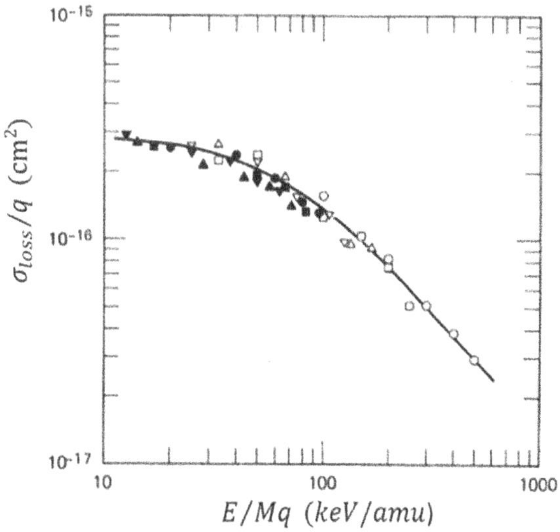

Fig. 4.2 CTMC cross sections σ_{loss} divided by the projectile charge q following A^{q+} + He collisions (q = 1–8), as a function of the collision energy in keV/amu divided by q. H^+, ○; He^{2+}, □; Li^{3+}, △; Be^{4+}, ▽; B^{5+}, ●; C^{6+}, ■; N^{7+}, ▲; O^{8+}, ▼; The full curve was obtained by fitting the theoretical function by an analytic function [1]

smaller than experimental one. Is appears evident that calculations and experiments were missing to give definitive conclusions.

The ability to determine total cross sections σ_{SI} and σ_{SC} using a simple classical electron independent model was helpful to plot the reduced cross section σ_{loss}/q, where $\sigma_{loss} = \sigma_{SI} + \sigma_{SC}$, as a function of the reduced projectile energy E/Mq, where M is the mass of the projectile (Fig. 4.2).

The theoretical data, represented for several projectile ions by empty and full objects, could be fitted by an analytical function (full red curve). This formula was found to agree well with quantum predictions at high energies, and to approach a limit at the lowest energies that is similar to that obtained previously from experiment.

CTMC method was improved nearly ten years later [2], by changing the initial conditions, in order to find the spatial and momentum variables for the case of general central interactions, and by modifying the expressions of the potentials $V(r)$ between the target nucleus and the active electron. The potential $V(r)$ has been represented either by an effective Coulombic potential or by a model potential

$$\begin{aligned} V_{eff}(r) &= -\frac{Z_{eff}}{r} \\ V_{mod}(r) &= -\frac{1}{r}\left[Z_T - 1 + \left(1 + Z_{eff}r\right)e^{-2Z_{eff}}\right] \end{aligned} \tag{4.2}$$

where $Z_T = 2$ in the present case, and $Z_{eff} = 1.685$.

Both potentials in absolute values are represented in Fig. 4.3 as a function of r. The potential $|V_{mod}(r)|$ is smaller than $\left|V_{eff}(r)\right|$ by a factor of about 1.65. This difference will play obviously a role on the force as seen by the electron and will modify the cross sections.

Fig. 4.3 Potentials $|V_{mod}(r)|$ (full red curve) and $|V_{eff}(r)|$ (dashed blue curve) as a function of the distance r between the target electron and the target nucleus, calculated using relations (4.2)

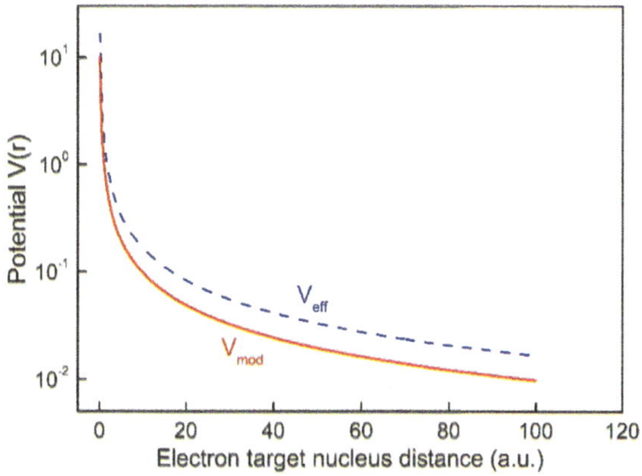

The radial distributions resulting from the use of the two potentials are compared with the corresponding quantum distribution (Fig. 4.4). The model potential gives rise to better results than the effective potential to predict the most probable value of the quantum distribution. The values for the total electron energy in the case of $V_{eff}(r)$ and $V_{mod}(r)$ are –1.71 a.u. and –2.84 a.u., respectively. The last value is in fair agreement with the real value of –2.904 a.u.

Figure 4.5 shows calculated cross sections for SI (top of the Figure), SC (middle) and DI (bottom), as a function of the projectile energy. They are compared with experimental results (full squares). Theoretical calculations using $V_{eff}(r)$ and $V_{mod}(r)$ are represented by red dashed and full curves, respectively.

The agreement between experiment and theory is good for SI, whatever the potential used. The calculated total cross section is not affected by the accuracy of the potential. For DI, the theoretical results are one order of magnitude larger than the experiment, even if the most sophisticated potential $V_{mod}(r)$ is introduced. According

Fig. 4.4 Calculated radial distributions using $V_{mod}(r)$ (red dashed curve), $V_{eff}(r)$ (red short-dashed curve) and quantum mechanics (full curve)

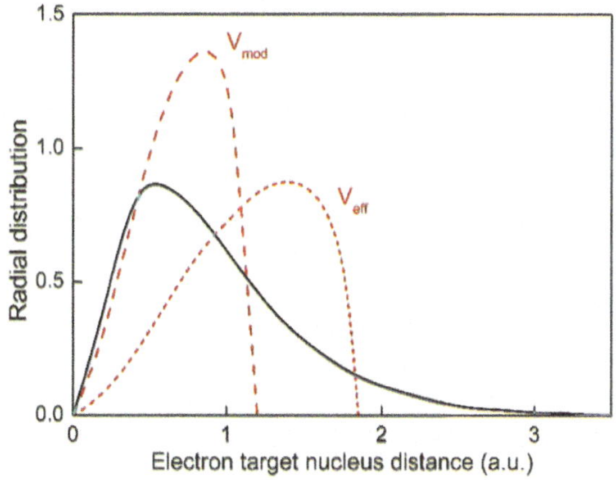

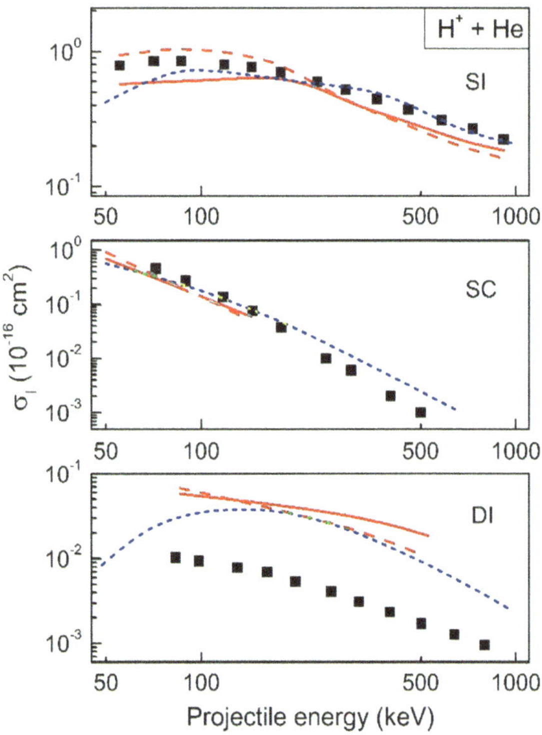

Fig. 4.5 Calculated CTMC cross sections following SI (top), SC (middle) and DI (bottom) in H^+ + He collisions as a function of the projectile energy. Experiment, full circles; Calculations: using $V_{eff}(r)$ and $V_{mod}(r)$ potentials, red dashed and full curves, respectively [2]; 3-body calculations using a phenomenological potential, blue short-dashed curves [3]

to the authors, this mismatch is probably due to the assumption of the independent electron model.

Improvement was made a few years later by adding a phenomenological potential $V_H(r)$ (cf. Chap. 2), in order to mimic the Heisenberg principle, but without the inclusion of electron correlation [3]. The results are presented in Fig. 4.5 for the three processes (blue short dashed curves). Compared to the previous calculations, the improvement is clearly visible at high energies for SI and SC. At the lowest energies, for SI, cross sections are smaller than experiment by a factor of less than two. DI cross sections are not better with the present model, confirming the conclusion of the previous authors who suggested that the inclusion of electron correlation would be necessary to improve the results.

Using a potential for the electron correlation that only includes the radial part (cf Chap. 2), DI cross sections have been revisited and compared to experiment and other previous theories [4] (Fig. 4.6).

The experimental data (full circles, empty circles and squares) are coherent with each other. The data represented by empty triangles are much larger than the previous data, because the cross sections include the effect of transfer ionization. The maximum of the experimental cross sections is observed for an energy of ~80 keV,

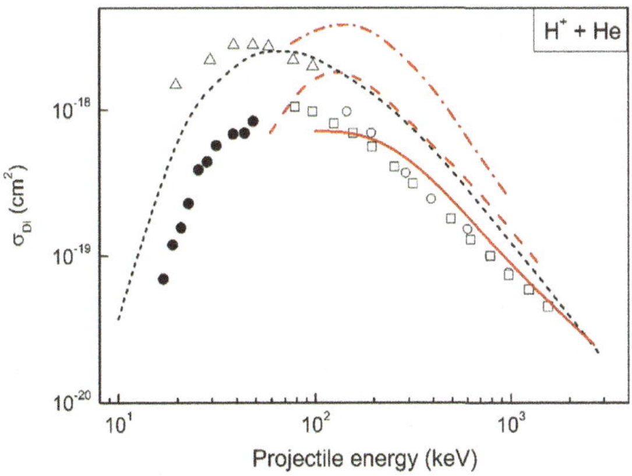

Fig. 4.6 Experimental (full and empty objects) and theoretical (curves) DI cross sections [4] following H^+ + He collisions as a function of the projectile energy. Binary encounter model, short dashed curve; red dashed curve, three-body CTMC calculation, red dotted-dashed curve [3]; CTMC calculation using radial correlation, red dashed curve [4]; four-body CTMC calculation, red full curve [5]

and is well reproduced, as well as the shape of the distribution, when using binary-encounter calculations (short dashed curve). With CTMC methods, the introduction of the radial electron correlation clearly improves the magnitude of the cross sections (red dashed curve) compared to the previous CTMC calculations (red dotted-dashed curve). Nevertheless, despite the present improvement, the present results are still higher than experiment. Two reasons may be invoked: Firstly, the initial radial electron distribution in the CTMC method has a maximum at small impact parameters, due to the classically forbidden region beyond the cut-off radius. Secondly, the lack of angular electron correlation may also lead to an increase of the DI probability, since both electrons may be on the same side of the target nucleus.

In 2002, the full electron correlation was included in the CTMC calculations [5]. Now, as seen in Fig. 4.6 (red full curve), the agreement is very good, except at energies smaller than 200 keV where calculated cross sections seem to be smaller than experiment. More effort would be devoted to calculating DI cross sections at lower energies. This effort is all the more desirable as SI cross sections have already been calculated in the energy range 5 keV–200 keV [6]. SI cross sections are plotted in Fig. 4.7 for the H^+ + He collision at energies smaller 200 keV. Experimental cross sections (full circles and squares) are compared with three-body (red dashed curve) and four-body (red full curve) CTMC calculations. Surprisingly, the best agreement is observed when electron correlation is not taken into account, since four-body calculations are smaller than experiment by a factor of ~2 at low energies. These results show that the introduction of electron correlation in CTMC calculations needs to be improved, and that calculations have to be extended to lower energies.

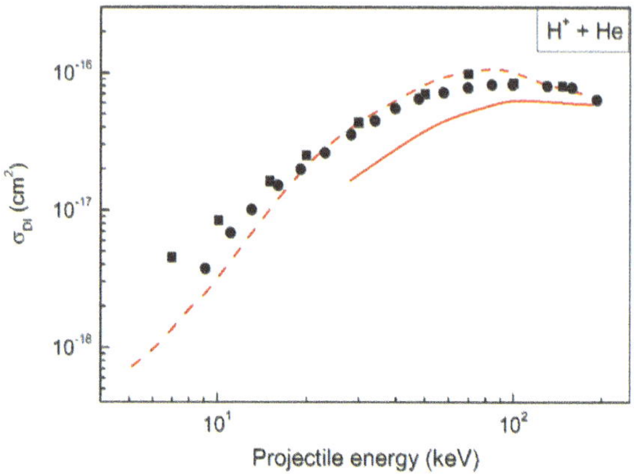

Fig. 4.7 Experimental (full circles and squares) and classical calculations (red curves), as a function of the projectile energy, for the collisions system H^+ + He, using an independent electron model (dashed curve) and a model that includes electron correlation (full curve). For more detail, see [6] and references therein

4.3.2 Partial Cross Sections

4.3.2.1 Electron Capture

The first classical study of partial cross section following electron capture in A^{q+} + He collisions was performed at intermediate and high projectile energies for a large number of values of q [3]. The electrons were treated independently. The total momentum distribution was divided into two independent parts and was chosen to fit the quantum one. In addition, the angular momentum was chosen by selecting an eccentricity for the Bohr orbit in the range from 0 to 1, allowing each electron to properly sample the angular momentum distribution and to limit the total angular momentum to 1.

The goal of the calculations was to verify the following hypothesis that was suggested after making experiments: The n level distributions are peaked at higher n values for SC than for TI [7]. Calculations were applied to $P^{15+} + He$ collisions at 100 keV/amu and compared with experimental cross sections after normalization (Fig. 4.8). Theoretical cross sections for SC (dashed curve) and DC (full curve) agree very well with experimental cross sections (full circles and squares).

The SC distribution is seen to be much broader and peaked at higher n value than the DC distribution. Unfortunately, at this projectile energy, no signature for TI was visible. Thus, calculations were performed at higher energies (250 keV/amu), at which TI is expected to be present. Figure 4.9 presents SC (full curve), DC (dashed curve) and TI (dashed-dotted curve) distributions, after normalization of the maximum of the cross sections, and without any comparison with experimental data that are not available. The present distributions for SC and DC ressemble those observed on Fig. 4.8. The TI and DC distributions are very simiar in shape. This similarity suggests that both proceed through a similar intermediate state which evolves into the separate final states. Consequently, SC and DC processes occur via

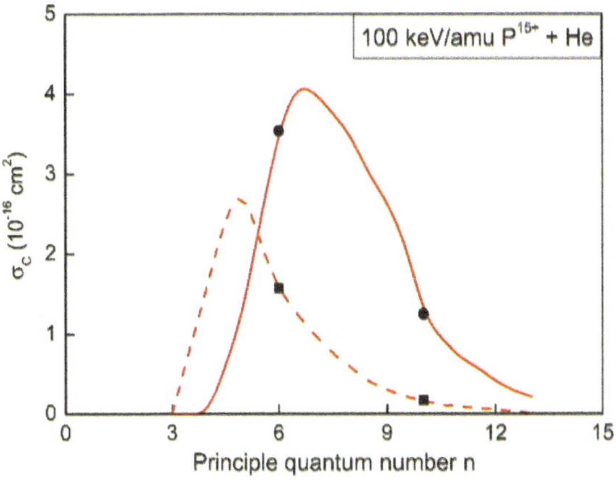

Fig. 4.8 n level distributions in 100 keV/amu P^{15+} + He collisions [3]. Experimental results: full circles and squares, SC and DC cross sections, respectively. Theoretical results: full and dashed curves, SC and DC cross sections, respectively

Fig. 4.9 n level distributions in 250 keV/amu P^{15+} + He collisions [3]. Theoretical results: full, dashed and dotted-dashed curves, SC, DC and TI cross sections, respectively

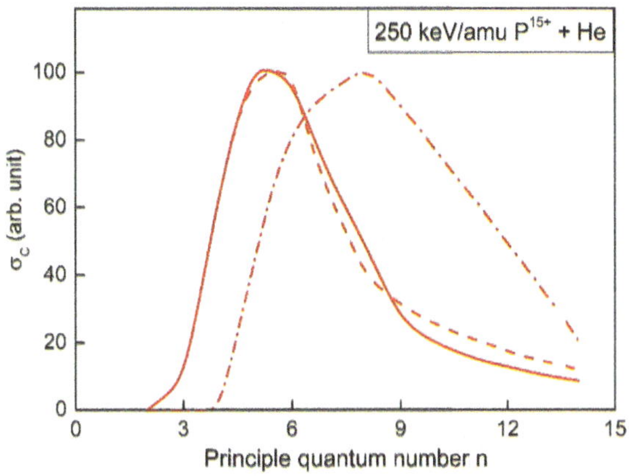

different routes, probably due, as mentioned by the authors, to the effect of electron correlation.

The first systematic study of the classical determination of *nl* distributions in A^{q+} + He collisions was conducted in 1990, for many values of *q* in the intermediate projectile energy range [8]. In principle, such a work is a first step that leads to an extraction of scaling laws, which have the great advantage to give any cross sections for any collision system. Unfortunately, it appears that, if scaling laws are available for total cross sections, they are quite impossible to determine since many parameters are involved, such as *n*, *l*, the projectile charge, and velocity.

In the work of 1990, CTMC calculations are compared with experiment or/and quantum calculations, when available. A benchmark was made for electron capture

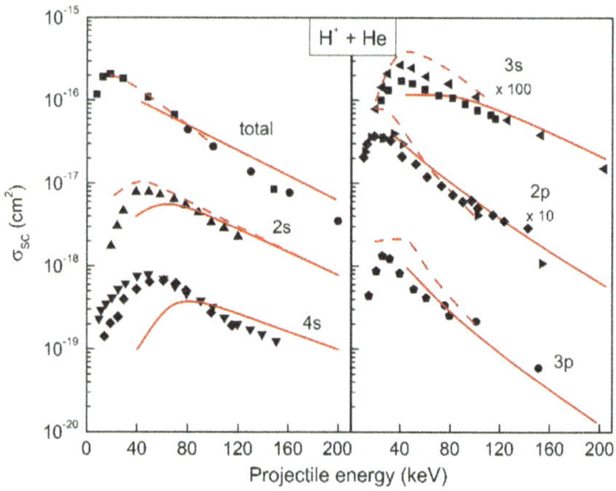

Fig. 4.10 Total and partial SC cross sections following H^+ + He collisions at projectile energies in the range from 20 to 200 keV (see [8] and references therein). Experiment, full objects; Theory: coupled-channel calculations, dashed curves; CTMC calculations, full curves

into different (n,l) states in H^+ + He collisions. Total and partial cross section for SC are presented in Fig. 4.10. CTMC calculations agree well with experimental data for $2l$ and $3l$ sates. Deviations in the shape of the cross sections are observed for the 4s state. Similar deviation was also present in $H^+ + H_2$ collisions.

An extension of CTMC calculations for charges q ranging from 1 to 10 was also performed (Fig. 4.11) at a projectile energy of 100 keV/amu [8]. The calculated CTMC cross sections agree well with experimental data for $q = 1$ and 10, but it is clear that many experimental results are missing to give a definitive conclusion on the ability for the classical calculations to reproduce the partial cross sections. In particular, even at this energy, experimental cross sections for higher values of n are desirable, and an extension to lower energies is required, since SC is dominant at the latter energies.

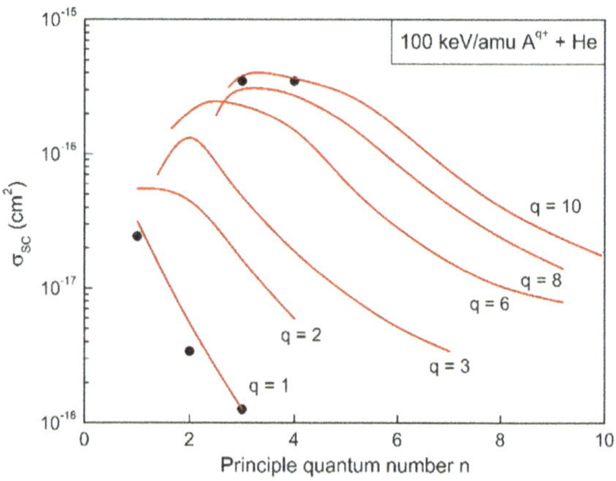

Fig. 4.11 n distributions following SC in A^{q+} + He collisions at a projectile energy of 100 keV/amu. Experiment, full objects for $q = 1$ and 10; CTMC calculations, red curves

4.3.2.2 Target and Projectile Excitation

Data on excitation of atoms by multicharged ion impact are, in contrast with electron impact, very limited in the case of total cross sections. Charge-scaling cross sections have been derived from experimental data using a variety of multicharged ions with charges up to 44 at intermediate projectile energies. Theoretical studies of *He* excitation were performed mainly using quantum mechanics. Classical mechanics is in principle a useful and good tool but is difficult to introduce for small values of the principal quantum number n of the excited state.

Despite the lack of comparisons between experiment and CTMC models, a few works has been devoted to calculation of partial cross sections on different excited states [9, 10].

We first focus on the particular collision system H + He that has retained attention and has been studied very recently in order to highlight the role of electron correlation during the collision [11]. Partial excitation cross sections on states with $n = 2$ were calculated in a large range of projectile energies (20 eV–100 keV). A multi-3 body CTMC (M3B-CTMC) model was first used, where the three electrons are treated independently of each other. Then, a full 5 body (5B-CTMC) model was performed, in which electron correlation is included.

Figure 4.12 shows energy distributions of electrons in H atoms at projectile energies of 1 keV and 100 keV, in the frame of the two models.

When electrons are independently treated, no marked excitation is observed at 1 keV (a), in contrast with what happens at 100 keV (b), where excitation on $n \geq 2$, corresponding to electron enerrgies larger than –7 eV, is clearly seen. The situation is completely reversed when the electron correlation is included, showing that this interaction strongly modifies the result. At this point, there is no evidence that the interaction between the two electrons causes this change. It is more than likely that it is the interaction between the projectile electron and each target electron which is the cause of this drastic change. That would be necessary to demonstrate.

If H electrons can be excited, it is expected that He electrons can also be excited. If it is the case, the contribution of excited He atoms will has to be subtracted from the total excitation cross sections.

Figure 4.13 presents the total energy distributions of He electrons at projectile energies of 20 eV, 1 keV and 100 keV.

The structure observed at energies of about –95 eV originate from non-excited configurations of *He*. At the lowest energies, no excited electrons are visible, whereas electrons with electron larger than –80 eV, that are the signature of excitation, are seen at 100 keV.

Figure 4.14 summarizes experimental (empty objects) and theoretical (curves) cross sections σ_{2l} for SE onto $2l$ states following $H + He$ collisions at projectile energies in the range from 10 eV up to 100 keV. On the experimental point of view, cross sections are found to reach a maximum of ~10^{-16} cm^2 at about 1 keV. Several theoretical calculation results using quantum mechanics (dashed-dotted and short dashed curves) deviate noticeably from experiment. The theoretical result shown by

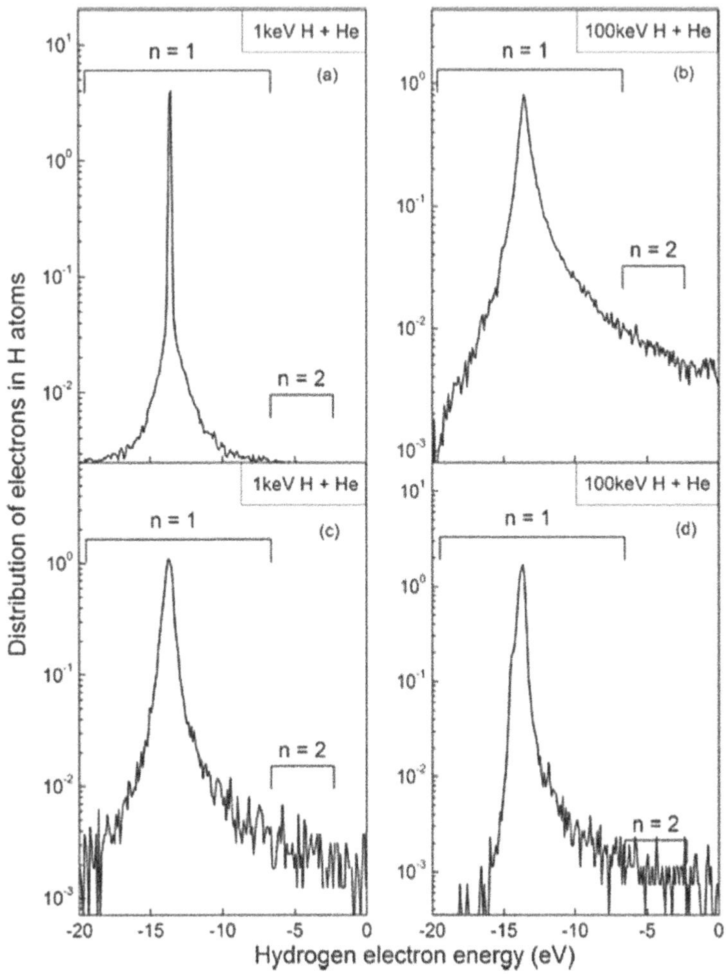

Fig. 4.12 Calculated energy distributions of the H atom electrons following H + He collisions, at projectile energies of 1 keV (left-hand side) and 100 keV (right-hand side). Calculations were performed in the frame of the classical independent electron model (top) and the classical correlated electron model (bottom). At electron energies higher than − 6.7 eV, H atoms are excited on *nl* orbitals ($n \geq 2$)

the dashed curves reproduces well the experiment. Finally, the very recent close-coupling calculations (black full curve) are found to be in reasonable agreement at energies lower than 1 keV, despite a discrepancy of up to a factor of 3. Part of this disagreement may originate from experiment itself since cross sections have been normalized to previous cross sections.

We focus on classical calculations (red curves). When the M3B-CTMC calculations is used (red short dashed curve), the theoretical calculations are in complete

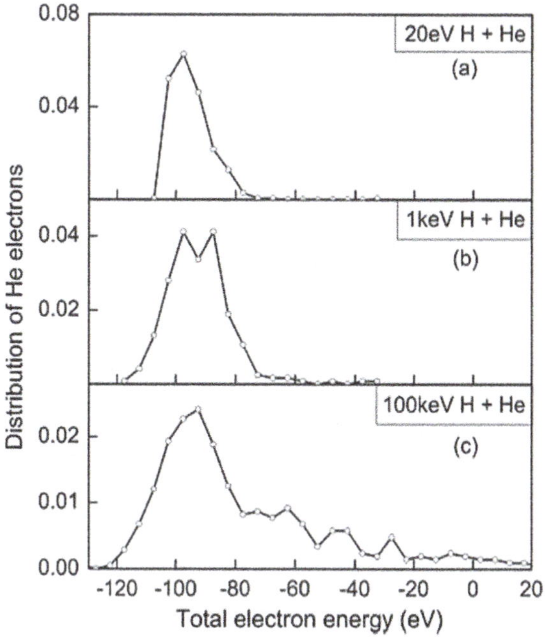

Fig. 4.13 Calculated energy distributions of He electrons following the reaction H(1 s) + He(1s^2) $\rightarrow$ H(nl) + He($n'l'n''l''$), with n, n' and $n'' \geq 2$, at projectile energies of 20 eV, 1 keV and 100 keV. The peak centered at about –95 eV corresponds to He atoms in the ground state. At 100 keV, a long energy tail is due to excited He atoms

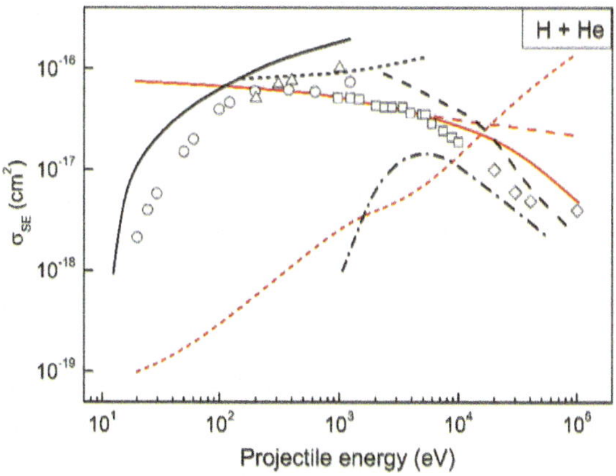

Fig. 4.14 SE cross sections following H + He collisions at projectile energies in the range from 10 eV up to 100 keV. Experiment, empty circles, triangles, squares and diamonds. Quantum calculations, black curves. Classical calculations: M3B-CTMC, red short dashed curve; 5B-CTMC, including *He* excitation, red dashed curve; 5B-CTMC, without *He* excitation, red full curve. For more details, see [11] and references therein

disagreement with the available experimental and theoretical results. The main reason is that the interaction between the projectile electron and the target electrons is not taken into account.

When the electron–electron interaction is included, the dependence of σ21 with Ep drastically changes compared to the previous result. When He is in its ground state or in excited states (dashed curve), σ_{2l} is in good agreement with experiment between 250 eV and 5 keV. However, our results strongly deviate from experiment outside this range. For example, at the highest projectile energy, our calculated cross section is larger than the experimental one by at least a factor of 5. At projectile energies smaller than 250 eV, σ_{2l} slightly increases when Ep decreases while the experimental results and the previous theory show a strong decrease.

However, when He remains in its ground state (solid curve), the agreement with experiment is good at energies larger than 200 eV. At the highest energies (higher than 5 keV), our results deviate from experiment only by a factor of less than 2. We have to recall that experimental results do not take into account the excitation of He. This point is crucial to understand the rather good agreement between experiment and our calculation.

4.4 Single and Multiple Differential Cross Sections

More than the calculation of the partial cross sections, the determination of differential cross sections in complex systems is decisive and remains a challenge to gain access to the most crucial details of the collision. Since the advent of multi coincidence detection in atomic physics with higher and higher precision, these details are now available and become a real benchmark to test quantum and classical theories.

Since the collision partners are considered as punctual particles, they are characterized, at the end of the collision, by their positions (spatial coordinates and/or angles), their velocities, and thus by their momenta and energies. Consequently, many quantities may be calculated, such as angular, energy or momentum differential cross sections, as well as combinations of these quantities.

In the case of CTMC calculations, the problem is the number of trajectories to introduce in order to obtain good statistics. For partial or total cross sections, ten thousand trajectories are enough. However, the number of trajectories has to be much larger (several billions) to extract DDCS or multiple differential cross sections (MDCS).

4.4.1 Single Differential Cross Sections

The first steps of theoretical calculations concern the determination of angular and energy distributions of the ejected electrons. The simplest system to test the validity of the various theories is again the H^+ + He collision system. Figure 4.15 presents, as

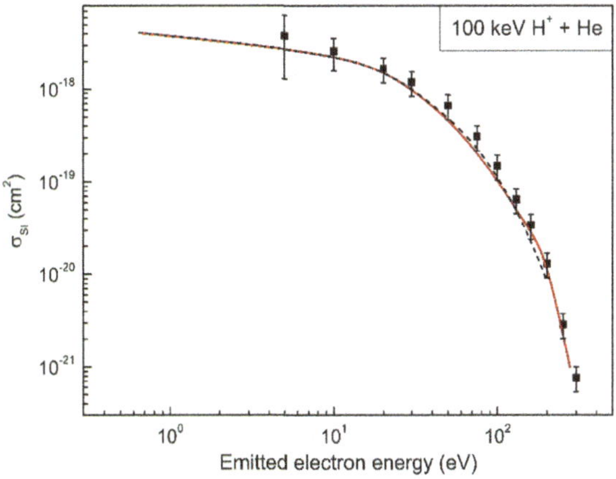

Fig. 4.15 Energy distributions of He electrons following single ionization in 100 keV H^+ + He collisions. Experiment, full squares; Born approximation calculations, dashed curve; 3B-CTMC calculations, red full curve

an example, the SDCS $d\sigma_{SI}/dE_e$, where σ_{SI} is the total ionization cross section and E_e is the energy of the ejected electron, as a function of E_e, for a projectile energy of 100 keV [12, 13].

The experimental cross sections, depicted by full squares, are compared with the Born approximation calculation (dashed curve) and 3B-CTMC calculations, using an effective charge of 1.69 for He and assuming each electron with a bounding energy of –0.904 a.u. The agreement between both theories and experiment is remarkable in the whole range of electron energies. It has to be noted that the agreement is also good at the projectile energy of 50 keV.

An example of angular emitted electron distributions is presented in Fig. 4.16 for the same system at 100 keV. The electron is emitted preferentially at small angles, since very small impact parameters are involved during the collision. The SDCS strongly decrease when the emission (or detection) angle θ_d increases. The agreement between the CTMC results (full curve) and experimental data (full squares

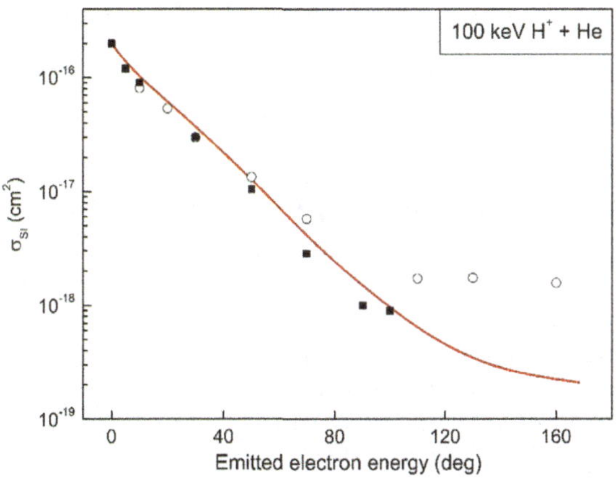

Fig. 4.16 Angular distributions of He electrons following single ionization in 100 keV H^+ + He collisions. Experiment, full squares and empty circles; 3B-CTMC calculations, red full curve

and open circles) is good for θ_d smaller than 90°. Strong deviations are observed at larger angles. However, it is difficult to really conclude since disagreement between experimental themselves are observed.

Since the development of cold targets in addition to coincidence measurements, transverse and parallel momentum distributions could be measured with a good precision. Coincidences between the projectile scattering angle were performed for collision systems with various perturbation strengths q/v_p, where v_p is the projectile velocity. In order to provide interpretation and predictions of scattering dynamics, 3B-CTMC method was developed, including electron-nucleus and nucleus-nuclaus interactions. This method was applied for various systems, such as 9.5 meV/amu Ni^{26+} + He or C^{6+} + He collisions for SI [14], or 3.6 meV/amu Se^{28+} + He collisions for SI and DI [15], with perturbation strengths from 1 to 5. To improve the electronic representation, model potentials derived from quantum mechanics were used. An example of cross sections $d\sigma_{SI}/dp_{//}$ is given in Fig. 4.17 for 9.5 meV/amu Ni^{26+} + He (q/v_p=1.33).

The agreement between experiment and theory is good, when multiplying theoretical cross sections by a factor of 2.41. In the case of electron distribution, even though the shape is well reproduced, a shift in momentum of about 0.2 a.u. is observed. Similar trends were also observed for 3.6 meV/amu Se^{28+} + He collisions. Following this good agreement, predictions were made for C^{6+} + He collisions at 506 keV/amu (q/v_p=1.33) and 165 keV/amu (q/v_p=2.34).

The $p_{//}$ distributions for electrons, recoil ions and change of projectile momentum are shown in Fig. 4.18. The results differ strongly from those observed in the previous case. At both energies, electron (full curves) and recoil ion (dashed curves) distributions are asymmetric. The projectile momentum change is found to significantly

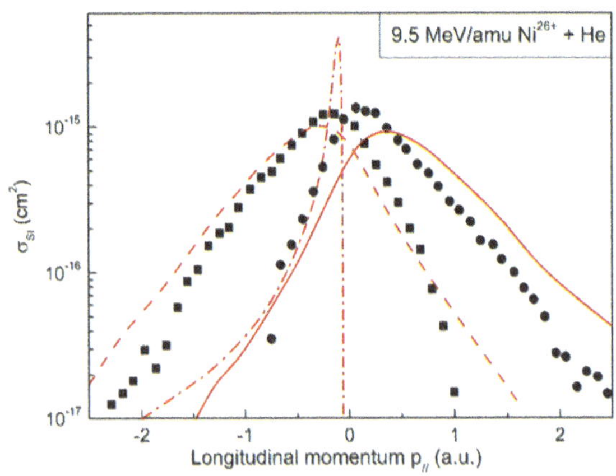

Fig. 4.17 Longitudinal momentum distributions for SI following 9.5 meV/amu Ni^{26+} + He collisions. Experimental electron and recoil ion spectra, full circles, and squares, respectively. 3B-CTMC calculations: electrons, full curve, dashed curve. The dotted-dashed curve represents the calculated change of longitudinal momentum of the projectile

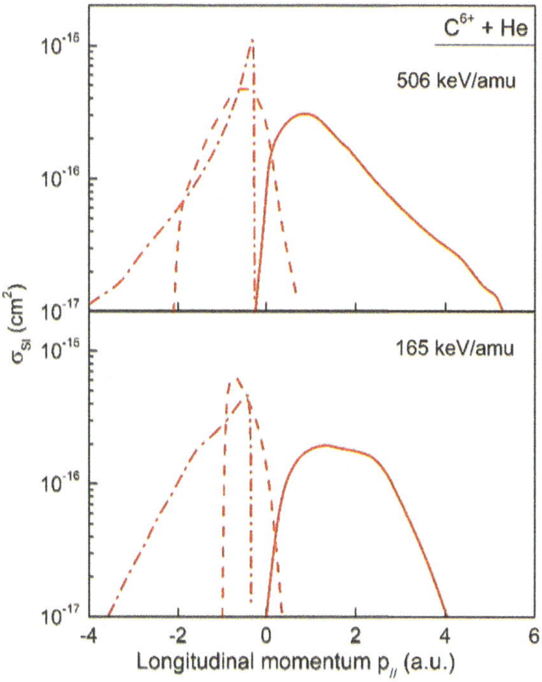

Fig. 4.18 Calculated longitudinal momentum distributions for SI following 506 keV/amu (top) and 165 keV/amu (bottom) C^{6+} + He collisions using a 3B-CTMC model. Electrons, full curve, recoil ions, dashed curve. The dotted-dashed curve represents the change of longitudinal momentum of the projectile

influence the three-body spectra and has a width that approaches and even exceeds that of the recoil ion, at energies of 506 and 165 keV/amu, respectively. According to the authors, this strong difference is mainly due to electron capture, which is not appreciable in the case of *Ni* and *Se* projectiles.

Finally, longitudinal electron spectra were calculated for C^{6+} + He collisions at 66, 48 and 34 keV/amu, corresponding to q/v_p values of 3.69, 4.33 and 5.14, respectively, and were compared with available experiments (Fig. 4.19) [14]. Even though the shape of the distributions are well reproduced, CTMC spectra are substantially shifted by a constant of +0.4 a.u. at each energy. This large difference was proposed to be explained by the double excitation channel. Indeed, this process would give rise to ~35 eV electrons ($p_{//} \sim$ 1.6a.u.). However, the corresponding cross sections is less than 10% of the total ionization cross section, so that it would not explain completely the present shift.

4.4.2 *Multiple Differential Cross Sections*

The determination of multi differential cross sections is the final and the most crucial step in the understanding of the mechanisms that are responsible for specific processes during the collision. One of the first studies was performed to analyze DDCS for the

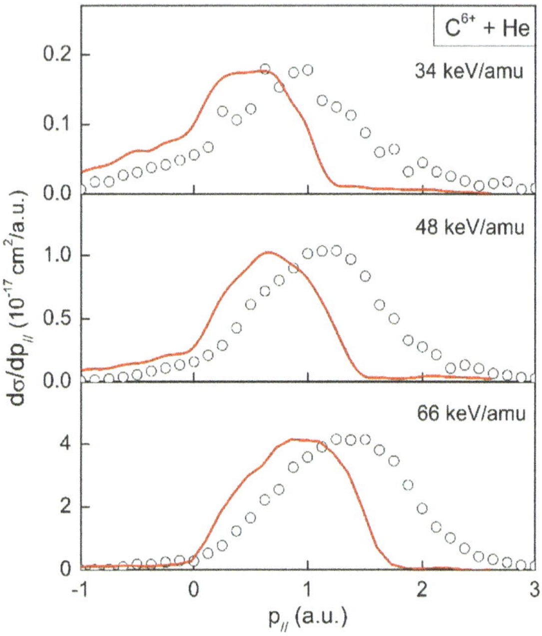

Fig. 4.19 Longitudinal electron momentum spectra for 34, 48, and 66 keV/amu C^{6+} + He collisions [14]. Experiment, empty circles; 3B CTMC calculations, red full curves

angular distribution of electrons following 100 keV H^+ + He collisions [12], for emitted electron energies of 13.6, 30, 54.4 and 100 eV, corresponding to velocities of 0.5 v_p, 0.74 v_p, v_p and 1.36 v_p, respectively.

The goal of the study was to get information in velocity space and in emission angle near $v_e = v_p/2$, at angles larger than 2.5°. At this electron velocity, a significant enhancement of the number of emitted electrons was expected. This enhancement, first observed for H^+ + H collisions, was attributed to the stranding of electrons on the transitory saddle region of the electric field between the target and the projectile nuclei. The experimental cross section $d^2\sigma_{SI}/dE_e d\Omega_e$, depicted in Fig. 4.20 (open circles), are compared with Plane-Wave Born Approximation (PWBA) calculations (dashed curves) and 3B-CTMC calculations (red full curves).The PWBA calculations reproduce quite well angular distributions for the two largest emitted electron energies, but are unable to describe them for E_e = 13.6 and 30 eV. The classical calculations agree rather well with experiment within 35% for θ_e smaller than 50°, with differences reaching a factor of 2 at larger angles, and for large E_e. For E_e = 54.4 eV, the CTMC calculations are able to enlighten the charge transfer to the continuum (CTC) process, that gives rise to a large intensity at small angles.

The same system was used to analyze the DDCS as a function of the emitted electron energy for various values of θ_e. This analysis was made to show that the CTC process peak is a classical phenomenon, and that it can predict both its shape and magnitude. In addition, the asymmetry of the peak, which is due to the residual target core, is also predicted by the CTMC calculations. These conclusions are summarized in Figs. 4.21 and 4.22 for θ_e = 1° and θ_e = 10, 50 and 90°, respectively. The

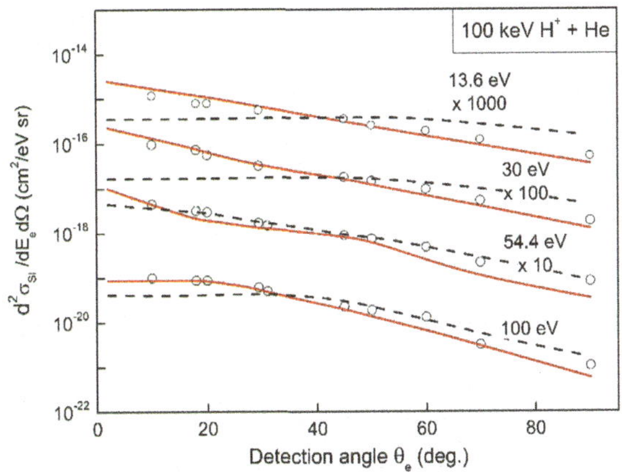

Fig. 4.20 Angular distributions $d^2\sigma_{SI}/dE_e d\Omega_e$ for SI following 100 keV/amu $H^+ + He$ collisions [12], for emitted electron energies of 13.6, 30, 54.4 and 100 eV, at an emitted angle of 1°. Experiment, empty circles; PWBA calculations, dashed curves; 3B-CTMC calculations, red full curves

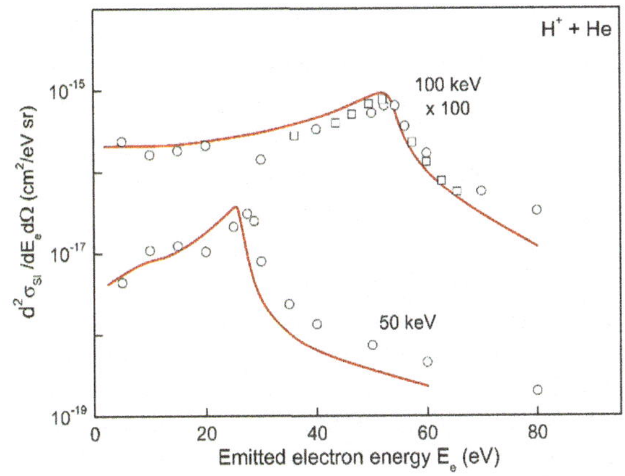

Fig. 4.21 DDCS for single ionization following $H^+ + He$ collisions at projectile energies of 50 and 100 keV/amu, as a function of the emitted electron energy, for a detection angle of 1°. Experiment, empty circles; 3B-CTMC calculations, full curves [13]

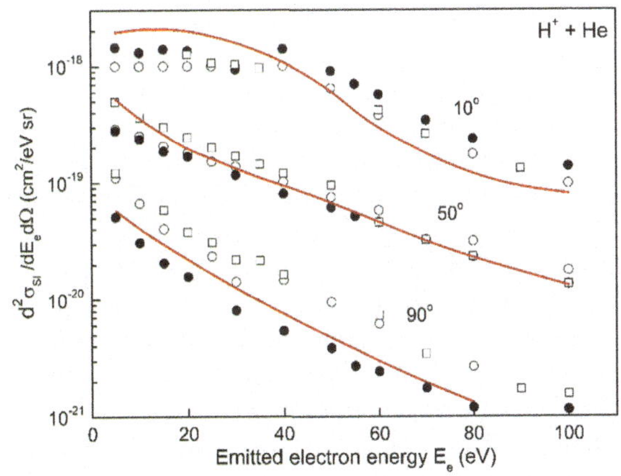

Fig. 4.22 DDCS for single ionization following $H^+ + He$ collisions at a projectile energy of 100 keV/amu, as a function of the emitted electron energy, for detection angles of 10, 50 and 90°. Experiment, empty and full circles, empty squares; 3B-CTMC calculations, full curves [13]

experimental cross sections (empty circles and square, full circles) are compared with the 3B-CTMC calculations (full curves). For $\theta_e = 1°$, the agreement is excellent. At larger angles, the CTMC calculations reproduce in average the experimental data. However, large differences are observed in experiment, especially at 90°, so that it is difficult to conclude about the validity of the classical model to give absolute DDCS with good precision.

4.5 Extension of the CTMC Model to Specific Problems

4.5.1 *Autoionization Following Double Electron Capture at Low Velocities*

It has been clearly demonstrated that several quantum effects, such as quantum division of orbits using quantum numbers, can be replaced by equivalent or quasi equivalent classical effects. In principle, it is not obvious that pure quantum effects, such as spin or Auger effect, may have their classical counterpart. However, even though the spin does not have great importance in atomic collisions, the Auger effect and autoionization processes are decisive in the understanding of the collision.

As mentioned in Chap. 2, Sect. 2.5.3.5, in most of the cases, the collision itself and the autoionization process are separated, since it is impossible to treat both simultaneously. Using quantum mechanics, the problem to treat is a pure four-body problem, including two nuclei and two electrons. The inclusion of electron correlations remains a real challenge, so that only a few works were devoted to the treatment of collision and post collision. In addition, since autoionization follows an exponential decrease of the form $e^{-\Gamma_A t}$, where Γ_A is the mean lifetime, the electrons are emitted not only in the Coulombic field of the projectile, but also in that of the residual target. Hence, both collision and post collision processes are not independent and cannot be, in principle, separated. This is all the more true since the projectile velocity is small (typically <0.1 a.u.), because the collision time τ_C is of the same order of magnitude than the autoionization time τ_A ($\tau_A = 1/\Gamma_A$).

The CTMC method has been shown to be a good tool even when including electron correlation. Consequently, it was applied to qualify and quantify the autoionization process, by analyzing the collision system Ne^{10+} + He [16] at a projectile energy of 150 keV, which has been extensively studied experimentally and theoretically.

At low velocities (an energy of 150 keV corresponds to a velocity of 0.53 a.u.), the capture processes are dominant. SC and DC cross sections are of the order of 2 10^{-15} cm^2 and 4 10^{-16} cm^2, respectively. For DC, one has to dissociate non autoionizing double capture (NADC), in which both captured electrons remain onto the projectile, and autoionizing double capture (ADC), where a captured electron is ejected. We also have to remind that ADC and TI give rise to the same final states, so that, in principle, they cannot be distinguished. However, since SI is unlikely to occur at low velocities, the probability for TI is also small compared to capture processes.

To characterize the processes, a time dependence of all the processes probabilities was achieved. First, since the cross section σ_P for process P is defined at infinity, an intermediate quantity $\tilde{\sigma}_P$, whose dimension is the same that of a cross section, was introduced. It was defined as

$$\tilde{\sigma}_P = \int_0^{b_{max}} 2\pi b P(b, t_C) db \tag{4.3}$$

where $P(b, t_C)$ is the probability for the process P after a collision time t_C. Obviously, $\tilde{\sigma}_P$ tends towards σ_P when t_C tends towards infinity.

We supposed that the collision time was varying from $-t_o$ to t_o. If the electrons are captured by the projectile $\tilde{\sigma}_{ADC}$ increases slowly with time while $\tilde{\sigma}_{NADC}$ decreases, verifying the relation $\tilde{\sigma}_{ADC} + \tilde{\sigma}_{NADC} = \sigma_{ADC} + \sigma_{NADC}$, whatever the value of t_C. In contrast, $\tilde{\sigma}_{SC}$, $\tilde{\sigma}_{SI}$ and $\tilde{\sigma}_{TI}$ are expected to be constant at $t > t_o$. By fitting these quantities with a formula of the form $a_o + b_o e^{-\Gamma_P t}$, where a_o and b_o are adjustable parameters, and Γ_P is a quantity that represents an average width, it is expected that Γ_{SC}, Γ_{SI} and Γ_{TI} are much larger than Γ_{ADC} and Γ_{NADC}.

In Fig. 4.23 are represented the calculated results for time evolution of $\tilde{\sigma}_{SI}$. This quantity remains constant after integration time larger than 50 a.u. (see the inset in Fig. 4.23), meaning that the SI process occurs during the collision.

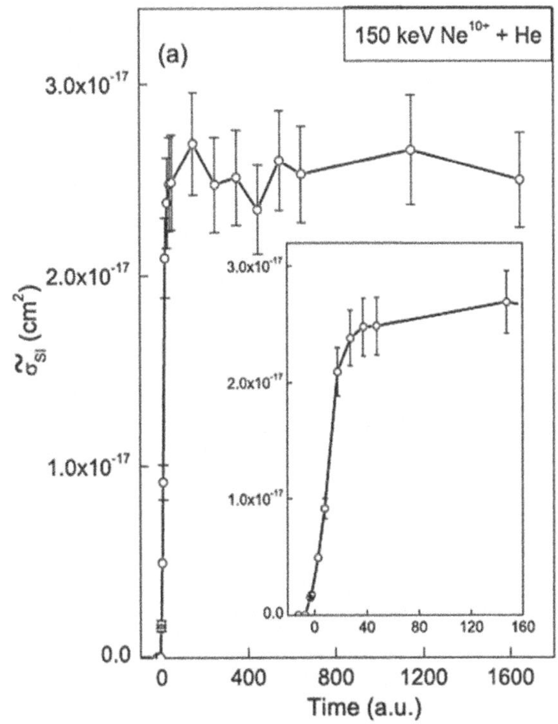

Fig. 4.23 Time evolution of the quantity $\tilde{\sigma}_{SI}$ using 4B-CTMC calculations, from $t = 0$ to 1600 a.u., in 150 keV Ne^{10+} + He collisions. Inset: zoom of time evolution for integration times smaller than 160 a.u

The same behavior was observed for the SC process. Finally, a fit of the cross sections gave rise to $\Gamma_{SI} = (0.10 \pm 0.05)$ a.u. and $\Gamma_{SC} = (0.063 \pm 0.007)$ a.u.

In contrast, as shown in Fig. 4.24, the quantity $\tilde{\sigma}_{NADC}$ decreases with time to the advantage of $\tilde{\sigma}_{ADC} + \tilde{\sigma}_{TI}$. Since the probability for TI is much smaller than that for ADC, and since $\tilde{\sigma}_{TI}$ is expected to be constant after a few a.u., the shape of the observed time evolution is mainly due to ADC. Using the fitting function, we found $\Gamma_{ADC} = (4.9 \pm 0.4)\ 10^{-3}$ a.u. and $\Gamma_{NADC} = (4.3 \pm 0.7)\ 10^{-3}$ a.u.

To verify the coherence of the latter values, an average autoionization width $\overline{\Gamma_A}$ was evaluated from previous experimental cross sections and calculations of individual widths. The value for $\overline{\Gamma_A}$ was found to be 4.4 10^{-3} a.u., which is surprisingly, even though it could be accidental, very close to that obtained theoretically. The agreement clearly demonstrated that 4B-CTMC calculations could describe the Auger effect without any ambiguity.

The integration time dependence of the cross sections is an important step in the analysis of the configurations that are populated during the collision. Indeed, it is

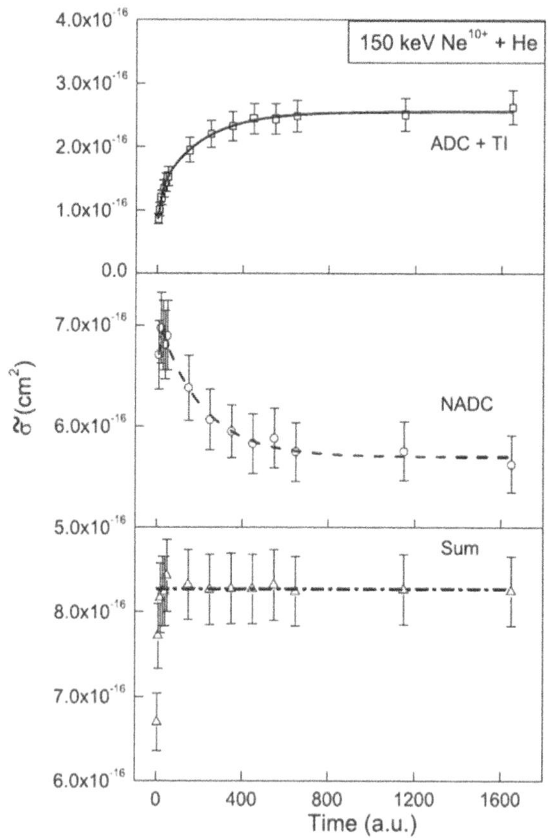

Fig. 4.24 Time evolution of the quantities $\tilde{\sigma}_{ADC} + \tilde{\sigma}_{TI}$ (top of the figure), $\tilde{\sigma}_{NADC}$ (middle) and the sum of all the quantities (bottom) using 4B-CTMC calculations, from $t = 0$ to 1600 a.u., in 150 keV Ne^{10+} + He collisions. For ADC + TI and NADC, a fit using the function $a_o + b_o e^{-\Gamma_P t}$ has been performed (full line on top and dashed for NADC) in order to determine an average autoionization width

well known for example that configurations of non-equivalent electrons (3, n), with $n > 5$, are mainly populated when the projectile velocity v_P decreases. It is thus expected that $\overline{\Gamma_A}$ would decrease with v_P.

4.5.2 *Single Ionization at High Velocities—Perpendicular Momentum Distributions*

4.5.2.1 Experimental and Theoretical Cross Sections

This last example will clearly show the powerful of the CTMC calculations, especially when the problem to treat seems to provide no solution. Since the early 2000's, much effort was devoted to the scattering of fast ionized projectiles in ion-atom collisions. Many experimental and theoretical studies were performed on the particular cases of 100 meV/amu C^{6+} + He and 3.6 meV/amu Au^{53+} + He collisions, for which the perturbation parameter $\eta = Z_P/v_P$ varies strongly [17–22]. The motivation was induced by the difficulty to treat theoretically the full problem for highly charged ions at intermediate energies, due to the strong interaction between the emitted electron and the outgoing projectile. For both collisions, the authors focused on the DDCS $d^2\sigma_{SI}/dq_\perp dE_e$, where $\overrightarrow{q_\perp}$ is the transverse momentum transfer of the projectile. Experimentally, whatever the value of E_e, the DDCS strongly decrease when increasing $q_\perp$, i.e. when decreasing the impact parameter b. At E_e larger than 50 eV, a peak appears at $q_\perp$ equal to the momentum of the emitted electron, showing that binary collisions between the projectile and the target electron plays a decisive role.

In the case of C^{6+} + He collisions, the experimental DDCS were nicely reproduced by Born and CDW-EIS calculations, as well as 3B-CTMC calculations. However, for Au^{53+} + He collisions, the agreement between theory and experiment was found to be poor. In Fig. 4.25, the DDCS for SI are represented as a function of $q_\perp$ and for E_e of 10, 50, 90 and 130 eV. The experimental cross sections exhibit structures centered at $q_\perp$ larger than 1 a.u. 3B-CTMC calculations (dashed curves) and quantum calculations (short dashed and dashed-dotted curves) fail to reproduce the experimental DDCS. After inclusion of electron correlation in CTMC calculations (red curves), 4B-CTMC results were improved, except at large values of $q_\perp$.

At this point, it was necessary to ask why no theory was satisfactory, both qualitatively and quantitatively, especially because the theories worked for the system C^{6+} + He. Aside from the difference in the parameter η between the two systems, the only notable difference is that the carbon ion has no electrons, while the Au ion has 26 electrons. It is therefore plausible that the electrons have a role to play during the collision, even if Au electrons are strongly bounded on the nucleus.

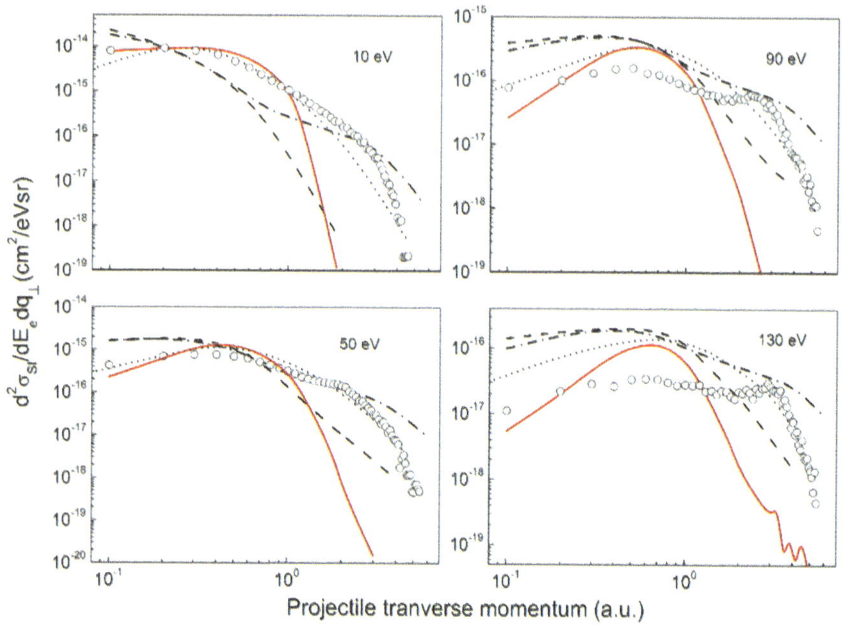

Fig. 4.25 DDCS as a function of the projectile transverse momentum following 3.6 meV/amu Au^{53+} + He collisions, for emission electron energies E_e = 10, 50, 90 and 130 eV. Experiment, open circles [17]; CTMC calculations using independent electron model, dashed curves [18]; quantum calculations, short-dashed curves [17, 19]; present 4B-CTMC model, convoluted with the experimental resolution of ~0.2 a.u., full curve

4.5.2.2 Theoretical DDCS Using Electron Projectiles

The DDCS for SI following 1.96 keV e^- + He collisions are presented in Fig. 4.26 as a function of $q_\perp$ and for the same values of E_e as previously used. The energy of 1.96 keV ensures that the electron and the Au ion have the same velocity. Two calculations were made. The dashed curves represent the DDCS using free electrons, i.e. electrons that are not bounded to the Au projectile. In this case, the electron-He^{2+} interaction is taken into account, inducing scattering angles that can be large compared to Au^{53+} scattering angles. In contrast, the full curves are the results of the calculations using electrons that are not free, but so close to the Au^{53+} projectile that the electron and Au^{53+} scattering angles are of the same order of magnitude. The DDCS were thus obtained by removing the electron-He^{2+} interaction.

For both approximations, binary structures are observed but, surprisingly, their maxima are shifted to larger values of $q_\perp$ compared to the maxima observed experimentally when all the interactions are taken into account. When the electron-He^{2+} interaction is removed, the position of the peaks maxima is in good agreement with the experimental one.

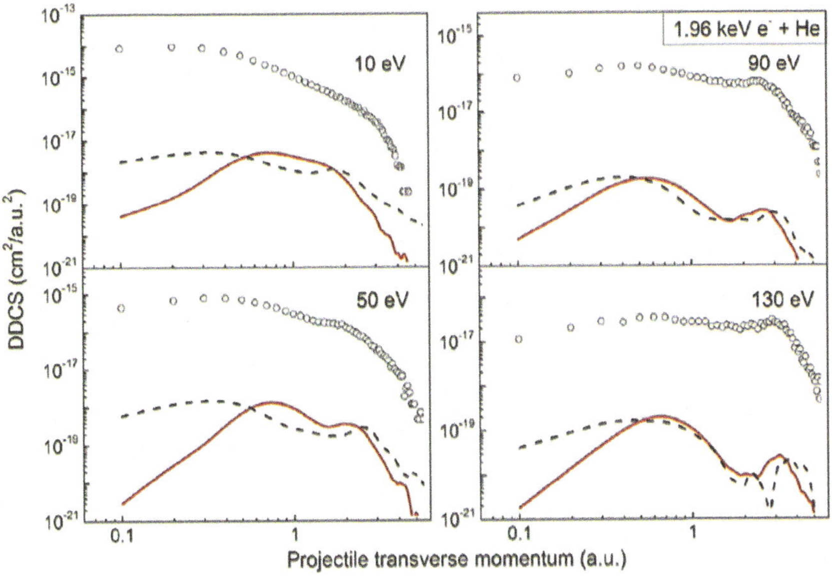

Fig. 4.26 DDCS as a function of the projectile transverse momentum following 1.96 keV e^- + He collisions, for emission electron energies $E_e = 10, 50, 90$ and 130 eV. Experiment, open circles [17]; 4B-CTMC calculations, dashed and full curves: classical calculation with the inclusion of the electron-He interaction, dashed curves; classical calculation after removing the electron-He interaction, full red curves

However, DDCS in the case of electrons as projectiles are at least two orders of magnitude smaller than the experimental DDCS. Consequently, one has to invoke the contribution of more than 100 electrons as projectiles order to quantitatively reproduce the binary contributions. The results are shown in Fig. 4.27. The full curve is the sum of both Au^{53+} (dashed curves) and e^- (short-dashed curves) contributions, weighted by coefficients that are used to fit the experimental data. As mentioned above, the number of electrons necessary to reproduce the experiment was of the order of a few hundred. It is important to note that the interaction between the electrons and Au^{53+} was removed, in order to simplify the calculations.

At 10 and 50 eV, the agreement between calculations and experiment was found to be rather good. The presence of the binary is well explained by the introduction of the electrons. At larger energies, the calculated binary contribution were smaller than the experimental one. At this point, the origin of such electrons had to be discussed. First, the huge amount of electrons was an indication that they could not originate from the Au^{53+} ions themselves. One has thus suggested that they would come from collisions between the projectiles and metallic surfaces, such as collimating slits, before penetrating in the collision zone and would follow the projectile in the same direction in with the same velocity. As shown previously [23] for 5 meV/amu ions penetrating carbon or insulator foils at normal incidence, the total number N of emitted electrons, for large values of the projectile charge q, is of the order of 0.3

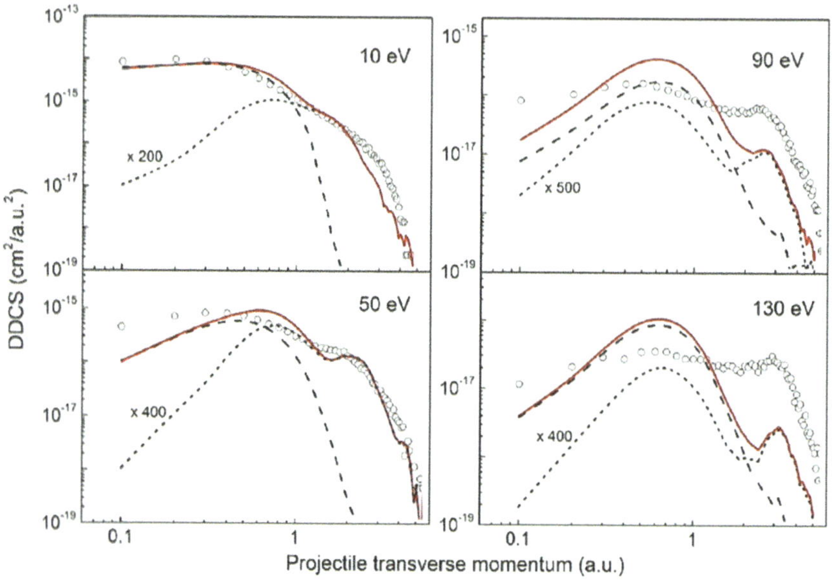

Fig. 4.27 DDCS as a function of the projectile transverse momentum following 3.6 meV/amu Au^{53+} + He and 1.96 keV e^- + He collisions, for emission electron energies $E_e = 10, 50, 90$ and 130 eV

q^2. Consequently, for C^{6+}, N ~ 10 while, for Au^{53+}, N ~ 1000, which is of the same order of magnitude as that found in the present calculations. However, the hypothesis of the presence of such electrons, also called convoy electrons, was rejected by the authors who published the experimental results. Two reasons were invoked. First, only a small fraction of the value of N have the same velocity and the same direction than that of the projectile. Secondly, the probability that they collide with the He target is negligible, due to the magnetic fields that strongly deviate the electrons. The last criticism would have been admissible if the electrons had been free. However, electrons interact not only with the magnetic field, but also with the Au ion. Hence, the interactions between electrons and Au on one side and electrons and the magnetic field on the other side compete.

To analyze the possible role of the convoy electrons, we assumed that they were randomly produced at distances in the range 0–50 a.u. At time $t = 0$, the distance r_e between an electron and the projectile is a few a.u. The time evolution of the r_e distribution is presented in Fig. 4.28 for a total number of 10,000 initial convoy electrons. Due to the presence of the magnetic field B (20 Gauss in the experiment), r_e increases with time. Nevertheless, even at $t = 10^8$ a.u., the distribution maximum is located between 10^{-6} and 10^{-2} cm, which is much smaller than the width of the target jet, so that most of the convoy electrons are able to reach the target, thus modifying the DDCS.

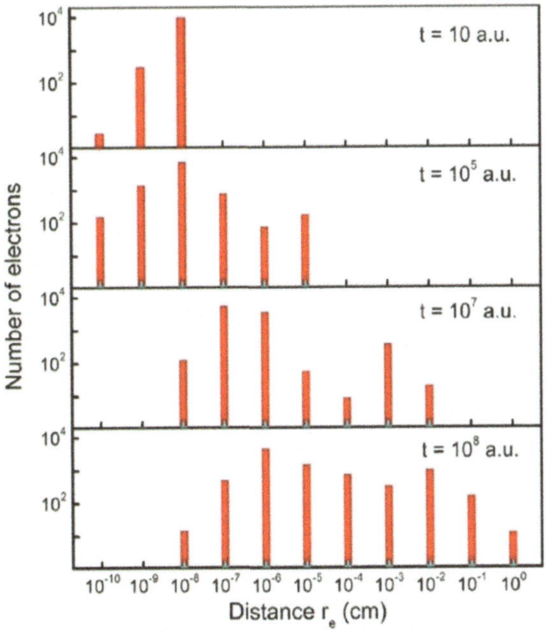

Fig. 4.28 Convoy electron distributions at different times of a collision. The electrons move in the field of an Au^{53+} ion and the magnetic field B of 20 Gauss used in the experiment. Time $t = 0$ corresponds to a situation where the electrons are located at distances re smaller than 10^{-7} cm. As time increases, B deviates the electrons and the average r_e distance increases. At the end of the collision (bottom of figure), most of the electrons can collide with the He target, since r_e is much smaller than 1 mm

In a second step, improvements of the previous model were performed. Instead of considering free convoy electrons, Coulombic interactions between these electrons and Au^{53+} were considered [24]. Nevertheless, to simplify the problem, the individual interactions between the electrons were neglected, and the presence of B was not considered in the Hamiltonian equations.

The DDCS and the TDCS were thus calculated and compared with the experiment and the previous theories. Figure 4.29 represents the DDCS when 200 convoy electrons were introduced. The present results were denoted by full red curves, while the dashed blue curves were the previous results of 4B-CTMC calculations when the convoy electrons were considered to be free.

At $E_e = 10$ and 50 eV, the presents results were in very good agreement with the experiment and the previous classical calculations, except for $q_\perp$ larger than 3 a.u. At $E_e = 90$ and 130 eV, the agreement with experiment was qualitatively reasonable, and the presents results were much better than those obtained with free electrons, despite discrepancies observed around $q_\perp = 1$ a.u.

4.5.2.3 Experimental and Theoretical TDCS

It is easily understandable that the TDCS measurements and calculations present a substantial challenge, as they can show up significant discrepancies which can be hidden in less differential cross sections. In the case of the Au^{53+} + He collisions, as

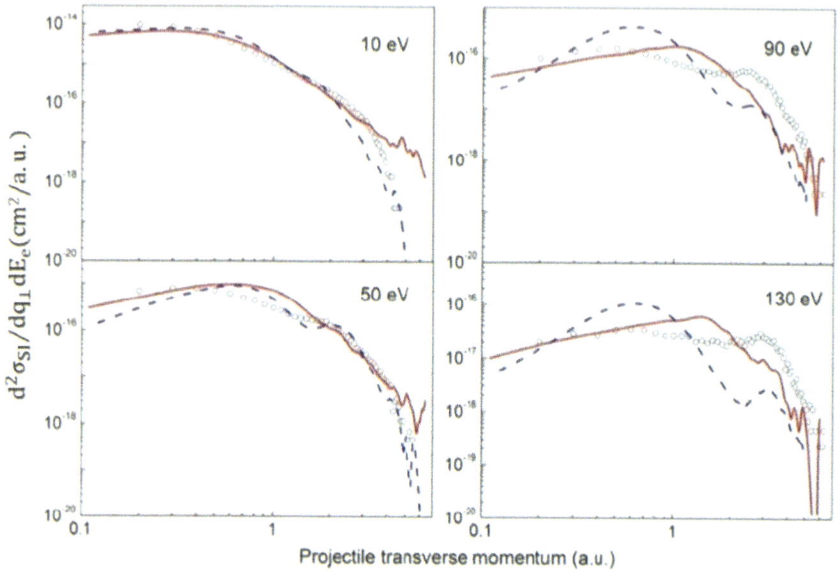

Fig. 4.29 DDCS as a function of the projectile transverse momentum following the 3.6 meV/amu Au^{53+} + He collisions, for the emission electron energies Ee = 10, 50, 90 and 130 eV. Open circles, experiment; the red full curves, the present MB-CTMC calculation with the inclusion of 200 convoy electrons that interact with the projectile; the blue dashed curves, recent four-body classical trajectory Monte Carlo (4B-CTMC) calculations [22] using 200 free electrons

mentioned in the introduction, none of the theoretical interpretations of the experimental results were satisfactory. As shown in Fig. 4.30, the experimental TDCS for $E_e = 4$ eV and $q_\perp = 0.65$ a.u. (open circles and black line) presented a maximum at an emitted electron angle θ_e around 0°. Among all the previous calculations, only a few theoretical results and references inside) were shown (blue curves on top of Fig. 4.30).

First Born approximation, two-Coulomb wave model, three distorted-wave (3DW) and three distorted-wave eikonal-initial-state (3DW-EIS) calculations gave rise to two structures, with a maximum at angles larger than 45°. On the contrary, the continuum-distorted-wave–eikonal-initial-state (3DW-EIS) calculation produced a wide structure whose maximum was located at about –50°. The results of the present calculations are presented on the bottom of Fig. 4.31, for 0, 10 and 20 convoy electrons (blue short dashed curve, blue dashed curve and red full curve, respectively). The region of integration for the angle was 10°. The electron yield from our calculations was normalized to that obtained experimentally. Without any convoy electrons, the structure (blue short curve) had a maximum at ~–50°, and no electron was emitted at angles larger than 100°. For 10 and 20 convoy electrons the structures were shifted to larger angles than those for no convoy electrons and presented a maximum at ~0°. It was also remarkable that the widths of the calculated structures reproduced quite well the experimental width. From this result, it appeared that the hypothesis

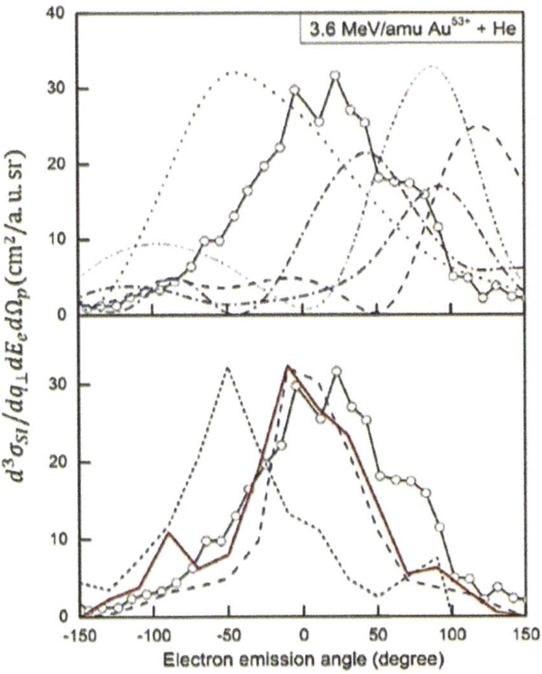

Fig. 4.30 Triply differential cross sections (TDCS) as a function of the emitted electron angle following the 3.6 meV/amu $Au^{53+} + He$ collisions, for $E_e = 4$ eV and $q_\perp = 0.65$ a.u. Open Circles, experiment; top: blue curves, previous theoretical calculations (see [22] and references therein); bottom: blue short-dashed curve, blue dashed curve and red full curve, the present MB-CTMC calculation with 0, 10 and 20 convoy electrons, respectively

of the presence of convoy electrons that actively participated to the collision can be considered as justifiable.

To confirm the validity of the hypothesis similarly calculated TDCSs were extracted using 20 convoy electrons (red curves in Fig. 4.31) for $E_e = 4$ eV and $q_\perp = 1$ a.u. (Fig. 4.31a), $E_e = 10$ eV and $q_\perp = 0.65$ a.u. (Fig. 4.31b) and $E_e = 10$ eV and $q_\perp = 1$ a.u. (Fig. 4.31c), for which experimental TDCSs were available (open circles).

Again, the calculated structures, in contrast with previous calculations, were very close to the experimental results. The maxima were peaked at ~0° and the widths were very close to those observed experimentally. For small values of emitted electron energies and relatively large values of $q_\perp$, the number of convoy electrons was of the order of 20, while the total number of convoy electrons was of the order of 200. As mentioned earlier, this was due to the fact that a small energy of the emitted electron energy corresponds to relatively large impact parameters, so that only the convoy electrons that were located far from the projectile axis contributed to the *He* ionization.

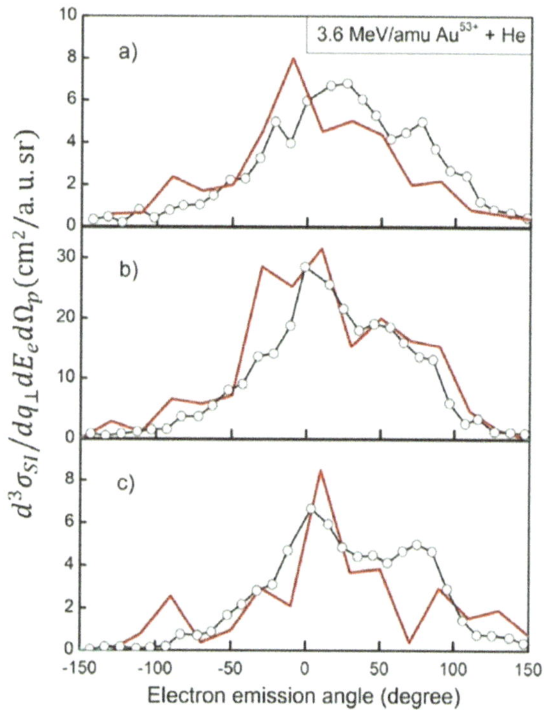

Fig. 4.31 TDCS as a function of the emitted electron angle following the 3.6 meV/amu Au^{53+} + He collisions, for **a** $E_e = 4$ eV and $q_\perp = 1$ a.u., **b** $E_e =$ 10 eV and $q_\perp = 0.65$ a.u., **c** $E_e = 10$ eV and $q_\perp = 1$ a.u. Open circles, experiment; red curves, present calculations including 20 convoy electrons

4.5.3 Extension to Fully Differential Cross Sections

Finally, very recently, the performance of the CTMC method was used to investigate for the ion-induced SI of *He* [25]. In this work, a Wigner type CTMC model was used in which the quantum–mechanical position distribution was approximated by a weighted integral of the microcanonical distribution of the binding energy of the electron. Fully differential cross sections (FDCS) were calculated for SI of He by 1 and 3 meV H^+ and 100 meV/amu C^{6+} projectile ions. These calculations completed quantum calculations which found deviations of the maximum in the energy distributions in the case of the binary peak. Here, the advantadge of the CTMC method was that it is a non perturbative theory that provides an exact description of the full dynamics of the multibody processes. Using different model potentials, the FDCS $d^5\sigma_{SI}/dq_\perp d\Omega_e d\Omega_p$, where Ω_e and Ω_p are the solid angles assiociated with the emitted electron and the scattered projectile, respectively.

Figure 4.32 presents classical FDCS as a function of the polar angle of the emitted electron for $E_e = 6.5 \pm 3.5$ eV, $q = 0.75$ a.u., in the scattering plane, allowing deviation with $\Delta\varphi_e = \pm\, 5°$, using different approximations.

When using traditional microcanonical CTMC calculations (blue short dashed curve), or improved CTMC calculations using hydrogenlike wave functions (blue dashed-dotted curve), the recoil peak, centered around 270°, is not present. In

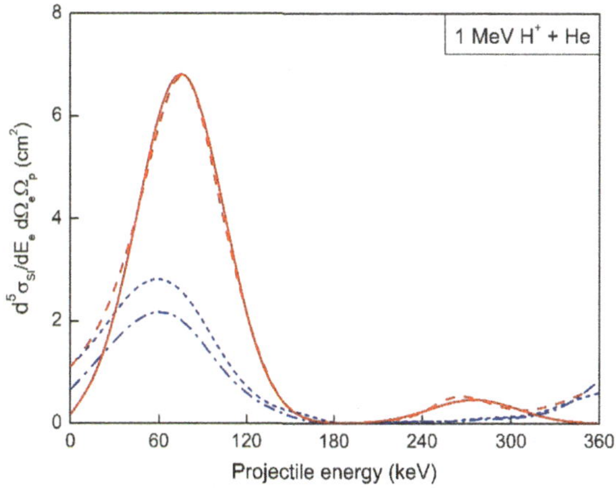

Fig. 4.32 FDCSs for the electron emission with $q = 0.75$ a.u. and $E_e = 6.5$ eV following the ionization of the helium atom by 1 meV protons in the scattering plane, as a function of θ_e. The data are results of the present CTMC calculations using different approximations (for more details, see text)

contrast, with more realistic wave functions (red dashed and full curves), the recoil peak appears. These wave functions are based on Roothann-Hartree–Fock (RHF) and Silverman-Platas-Matsen (SPM) position distributions [25]. In the following only the CTMC calculations using SPM wave functions are compared with experiment and quantum calculations.

In Fig. 4.33, for 1 meV H^+ + He collisions, experimental FDCS (open circles) are compared with quantum calculations (blue dashed, short dashed and dashed-dotted curves) and CTMC-SPM calculation (full red curve). Using quantum mechanics, the experimental recoil peak was well reproduced in shape, but not in intensity. In the case of the binary peak, the shape was also reproduced, but a shift of about 6.5° was found, compared to the experiment. The CTMC model provided a reasonable description of the experiment, except at angles close to 0°. A shift of the CTMC binary peak was also present but was found to be twice smaller than that found using

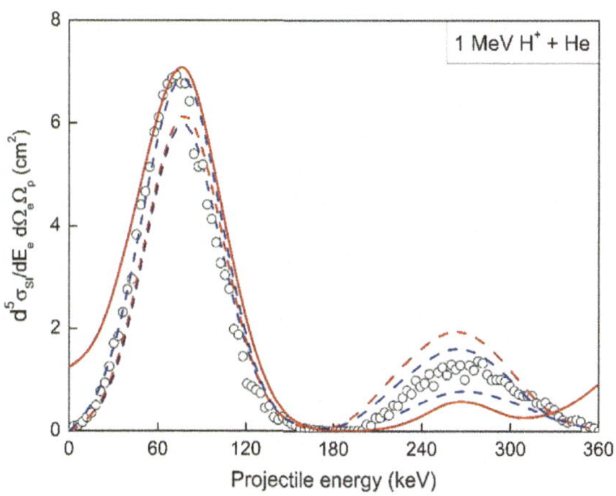

Fig. 4.33 Comparison of the FDCSs obtained by the present CTMC-SPM model (full red curve) for 1 meV proton impact in the scattering plane with the experimental data [25] (empty circles) and the predictions of quantum–mechanical models (blue curves)

quantum mechanics. The recoil peak calculated with the CTMC method was smaller and narrower than that observed in the experiment and predicted by all the quantum–mechanical models. Modifications of the model potential were made, but they did not improve the description of the recoil peak. The author concluded that quantum effects, such as the coherence properties of the projectile, may play a role. Similar conclusions may have been advanced by the author for the other studied collision systems.

4.6 Preliminary Conclusion

Through various examples, classical calculations have shown that, even in the finest details, they were able to give excellent results, in part thanks to the introduction of electron–electron interaction. The method turns out to be a precious tool in the description of four-body processes, whether in the calculation of total, partial or multiple differential cross sections.

In the following, more complex systems will be reviewed, involving more than two electrons.

References

1. Olson, R.E.: Electron-capture and impact-ionization cross sections for multiply charged ions colliding with helium. Phys. Rev. A **18**, 2464 (1978)
2. Reinhold, C.O., Falcon, C.A.: Classical ionization and charge-transfer cross sections for H^+ + He and H^+ + Li^+ collisions with considerations of model interactions. Phys. Rev. A **33**, 3856 (1986)
3. Wetmore, A.E., Olson, R.E.: Electron loss from helium atoms by collisions with fully stripped ions. Phys. Rev. A **38**, 5563 (1988)
4. Montemayor, V.J., Schiwietz, G.: Dynamical target screening for two-active-electron classical-trajectory Monte Carlo methods for H^++ He collisions. Phys. Rev. A **40**, 6223 (1989)
5. Morita, S., Matsuda, N., Toshima, N., Hino, K.: Ionization of stabilized helium atoms by proton and antiproton impacts. Phys. Rev. A **66**, 042719 (2002)
6. Dimitriou, K., Aumayr, F., Katsonis, K., Winter, H.P., Chibisov, M.I., Janev, R.K., Urbain, X., Brouillard, F.: Atomic data for H^+ + He($1s^2$), He (NLM) collisions: single ionization, excitation and charge exchange cross sections. In: Atomic and Plasma—material interaction data for fusion (Vienna, Austria), vol. 13, pp 88 (2007)
7. Datz, S., Bottcher, C., Andersen, L.H., Hvelplund, P., Knudsen, H.: Electron transfer processes in collisions of highly charged energetic (0.1–1.0 MeV/nucleon) ions with helium atoms. Nucl. Intr. Meth. Phys. Res. A **262**, 62 (1987)
8. Meng, L., Reinhold, C.O., Olson, R.E.: Subshell electron capture in collisions of fully stripped ions with He and H2 at intermediate energies. Phys. Rev. A **42**, 5286 (1990)
9. Reinhold, C.O., Falcon, C.A., Miraglia, J.E.: Ionization and excitation to Rydberg states in collisions of fully stripped ions with hydrogen atoms. J. Phys. B **20**, 3737 (1987)
10. Reinhold, C.O., Olson, R.E., Fritsch, W.: Excitation of atomic hydrogen by fully stripped ions. Phys. Rev. A **41**, 4837 (1990)
11. Frémont, F., Belyaev, A.K.: Excitation of hydrogen atoms in collisions with helium atoms: the role of electron-electron interaction. J. Phys. B **50**, 045201 (2017)

12. Olson, R.E., Gay, T.J., Berry, H.G., Hale, E.B., Irby, V.D.: Saddle-point electrons in ionizing ion-atom collisions. Phys. Rev. Lett. **59**, 36 (1987)
13. Reinhold, C.O., Olson, R.E.: Classical two-center effects in ejectd-electron spectra from p^+, p^-, and He^{2+} + He collisions at intermediate energies. Phys. Rev. A **39**, 3861 (1989)
14. Olson, R.E., Wood, C.J., Schmidt-Böcking, H., Moshammer, R., Ullrich, J.: Complete single-ionization spectra for strong perturbation collisions. Phys. Rev. A **58**, 270 (1998)
15. Moshammer, R., Ullrich, J., Kollmus, H., Schmitt, W., Unversaget, M., Jakutzki, O., Mergel, V., Schmidt-Böcking, H., Mann, R., Woods, C.J., Olson, R.E.: Double ionization of helium and neon for fast heavy-ion impact: correlated motion of electrons from bound to continuum states. Phys. Rev. Lett. **77**, 1242 (1997)
16. Frémont, F.: Quasiclassical treatment of the Auger effect in slow ion-atom collisions, Phys. Rev. A **96**, 032712 (2017)
17. Moshammer, R., Perumal, A.N., Schulz, M., Rodriguez, V.D. Kollmus, H., Mann, R., Hagmann, S., Ullrich, J.: Three-body Coulomb problem probed by mapping the Bethe surface in ionizing ion-atom collisions. Phys. Rev. Lett. **87**, 223201 (2001)
18. Olson, R.E., Fiol, J.: Mapping of the Bethe surface in single-ionization ion-atom collisions. J. Phys. B At. Mol. Phys. **34**, L625 (2001)
19. Rodriguez, V.D.: CDW-EIS theoretical calculations of projectile deflection for single ionization in highly charged ion-atom collisions. Nucl. Instrum. Methods Phys. Res. B **205**, 498 (2003)
20. Fiol, J., Olson, R.E.: Three- and four-body dynamics in fast heavy ion-atom ionization. J. Phys. B At. Mol. Phys. **37**, 3947 (2004)
21. Fainstein, P.D., Gulyás, L.: Three- and four-body dynamics in single ionization of He by swift highly charged Au^{53+} ions. J. Phys. B At. Mol. Phys. **38**, 317 (2005)
22. Frémont, F.: Transverse momentum transfer distributions following single ionization in 3.6 MeV/amu Au^{53+}+ He collsions: A 4-body classical treatment. Atoms **6**, 68 (2018)
23. Schiwietz, G., Xiao, G., Grande, P.L., Schmoldt, A., Grether, M., Köhrbrück, R., Spieler, A., Stettner, U.: Non linear effects in ion induced electron emission. Scanning Microsc. **12**, 265 (1998)
24. Frémont, F.: Doubly and triply deifferential cross sections for single ionization of He by fast Au^{53+}using a multi-body quasiclassical model. Atoms **8**, 19 (2020)
25. Sarkadi, L.: Fully differential cross sections for the single ionization of helium by fast ions: classical model calculations. Phys. Rev. A **97**, 042703 (2018)

Chapter 5
Extension of CTMC Calculations to Multielectron Systems

Abstract Due to the quality of a large number of results obtained with the CTMC, method for collisions involving H and He targets, it is obviously tempting to extend the method to multielectronic targets. The simplest target to use seems to be Li, which has two electrons in the 1s orbital and a less bound electron on the 2s shell. Then, the method is extended to multielectronic targets such as Ne or Ar.

5.1 Interest and Motivation

Due to the quality of a large number of results obtained with the CTMC method for collisions involving H and He targets, it is obviously tempting to extend the method to multielectronic targets. The simplest target to use seems to be Li, which has two electrons in the 1s orbital and a less bound electron on the 2s shell. Rather than reasoning like this, we will study the processes according to the energy of the active electrons. We are going to distinguish the targets whose active electrons are energetically little bounded to the nucleus in comparison with the other electrons. This is true for example for the alkaline elements (Li, Na, K, Cs), and for the atoms for which an electron is initially in a highly excited state. For these elements, apart the active electron, the other electrons can be, in first approximation, considered as frozen (Fig. 5.1).

In the case where the targets have active electrons are on similar shells, such as Ne, Ar, Kr or Xe, the problem is more complex, because electrons on a same orbital cannot be considered anymore as independent. Nevertheless, to calculate total cross sections, independent electron models (IEM) are efficient, as well as scaling law. One has to be careful, because even though scaling laws and IEM are really useful, they give rise to absolute results with an uncertainty that can be larger than 10%. This value is considerable when problems involving a huge number of particles have to be solved.

Before discussing results obtained with the CTMC method, two examples using scaling laws derived from classical results, and the Classical Over Barrier Model, will be given.

© The Author(s), under exclusive license to Springer Nature Switzerland AG 2021
F. Frémont, *Classical Treatment of Collisions Between Ions and Atoms or Molecules*, Springer Series on Atomic, Optical, and Plasma Physics 118, https://doi.org/10.1007/978-3-030-89428-3_5

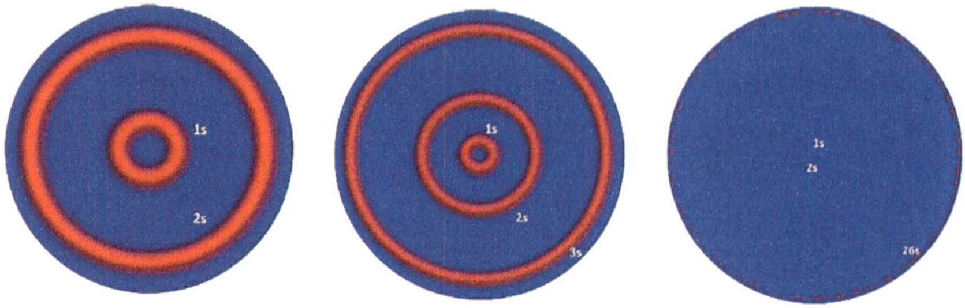

Fig. 5.1 Schematic $1/I_p$ distributions of Li ($1s^22s$), Na ($1s^22l^63s$) and Na ($1s^22l^626s$) (left side, middle and right side, respectively). Each outer shell electron can be considered as independent from the inner shell and core electrons

5.2 Simple Models for a Preliminary Study

5.2.1 Classical Over Barrier Model

A great effort has been devoted to the determination of total and partial cross sections following electron capture and then, ionization, using the COB model [1]. In principle, this model is valid at low velocities compared to that of the target active electron and has no projectile velocity dependence. Several years later, extensions of this simple one-dimension model have been added, taking into account collision parameters such as the projectile velocity.

Figures 5.2 and 5.3 show typical energy gain spectra of scattered projectile ions following SC and DC in $Xe^{q+} + Ar$ ($q = 7$–11) collisions, respectively, at a projectile energy of 9 keV [2].

For SC and DC, the experimental spectra are located at energies larger than 10 eV and 50 eV, respectively, whatever the value of q. The arrows, indicating the energy

Fig. 5.2 Experimental energy gain spectra (empty circles) following SC in 9 keV $Xe^{q+} + Ar$ ($q = 7 - 11$) collisions. The arrows indicate the mean gain energy obtained using the COB model

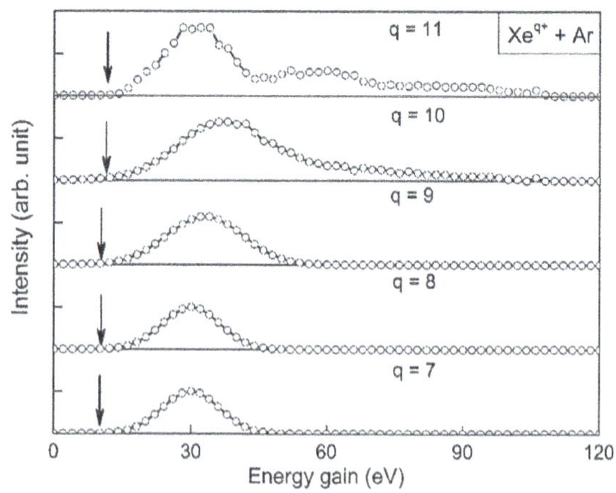

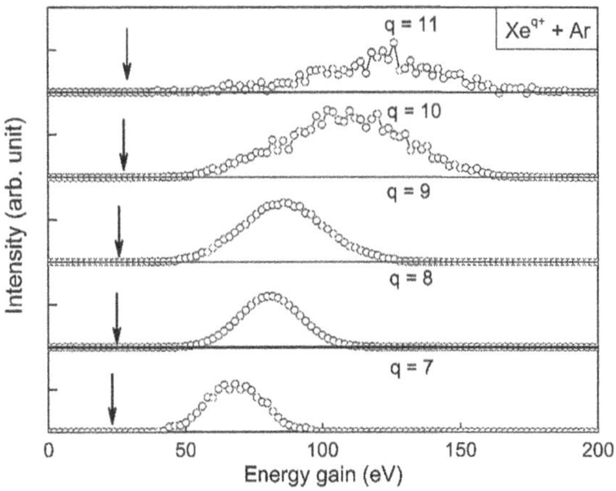

Fig. 5.3 Experimental energy gain spectra (empty circles) following DC in 9 keV $Xe^{q+} + Ar$ ($q = 7-11$) collisions. The arrows indicate the mean gain energy obtained using the COB model

gain estimated from the COB model are far below the experiment. It is thus clear that the COB model is not adapted to such collision systems. Several reasons may be responsible for these large discrepancies. First, the estimated energy gain depends strongly onto the target effective charge. Second, the product Ar ions can be in excited states during the collision, giving rise to a change of energy gain. Finally, for such systems, electron correlation may play a non-negligible role.

The COB model was strongly modified many years later in order to calculate either capture or multiple ionization cross sections, at low [3] and high [4] projectile energies E_p (corresponding to velocities v_p). At very low E_p (below 1 keV/amu), the total capture cross sections are no more independent onto E_p. To take into account E_p, the authors [3] approximated the interaction potential by the dipole potential $V_p(r) = -\alpha q^2/2r^4$, where α is the target polarizability, and r is the internuclear distance. Since the classical turning point r_c and the corresponding impact parameter b_c are related by

$$b_c = r_c\sqrt{1 - \frac{V_p(r_c)}{E_p}} \tag{5.1}$$

the capture cross section will depend onto E_p. The results for this dependence is presented in Fig. 5.4 for the collision system Kr^{9+} + Ne, in the case of capture of 1 up to 4 electrons. The calculated cross sections (full blue curves) agree well with experiment (empty circles) for the capture of 3 and 4 electrons, but discrepancies with experiment are revealed for SC and DC. Indeed, whereas experimental cross sections are nearly constant in the whole range of energies explored, calculations show a deep minimum at E_p of 1 eV and 2 eV, underestimating the experimental results. Even though this model has the advantage to give results very quickly, one has to note that fitting parameters where used in the present model. As a conclusion,

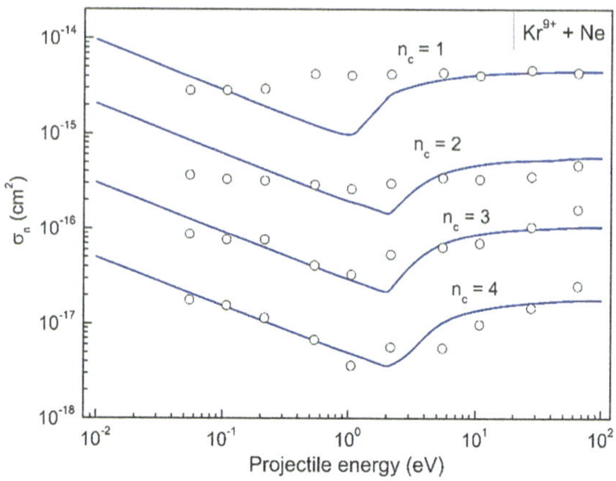

Fig. 5.4 Multiple capture cross sections σ_n ($n = 1 - 4$) following Kr^{9+} + Ne collisions as a function of the projectile energy [3]. Experiment, empty circles; modified COB model, full blue curves

it is difficult to imagine that such a model can be used to replace more sophisticated models.

At high energies, the COB model was also used for its simplicity, compared to quantum and CTMC calculations, especially when molecular targets are involved. The model was modified according to the following hypothesis [4]: the release and capture distances, R_r and R_c, respectively, are given by

$$\begin{aligned} R_r &= \frac{Z + 2\sqrt{qZ}}{I} \\ R_c &= \frac{2q}{v_p} \end{aligned} \tag{5.2}$$

If $R_r < R_c$, the electrons will be captured. If $R_r > R_c$, the electrons will be ionized when they will obtain enough kinetic energy to escape. Thus, probabilities for multiple ionization could be calculated. As an example, Fig. 5.5 shows the ratios R_{21} and R_{31} of double and triple to single ionization, respectively, following Cl^{q+} + Ne collisions ($q = 9$–13) at an impact energy of 34 meV. The COB results (full blue curves) are compared with the semi-classical approximation (SCA) calculation results and the available experimental data. Bothe calculations agree rather well with experiment, even though SCA results are underestimated, and COB results are overestimated. However, it is quite clear that the COB results are better than the SCA results.

The agreement is also good when the ratios R_{21} and R_{31} are represented for the collision system Cl^{12+} + Ne as a function of E_p (Fig. 5.6). As a conclusion, the author could be reasonably claim that this model provides a convenient and easy way to obtain multiple fractions of complex molecules.

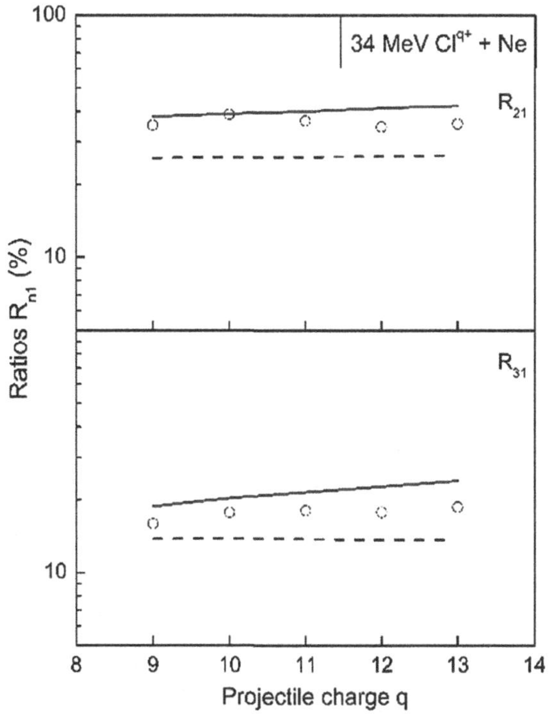

Fig. 5.5 Ratios R_{21} and R_{31} of double and triple to single capture cross section following 34 meV Cl^{q+} + Ne collisions ($q = 9$–13). Experiment, empty circles; dashed curves, SCA calculations; full blue curves, COB model

5.2.2 Scaling Law for Multiple Electron Capture

Many scaling laws were established and derived either from experiment or theories. The scaling laws are considered as a simple and rapid alternative to calculations involving complex systems or complex collisions partners. However, scaling laws were only determined for specific projectile velocity regimes, at low or at high velocities. Consequently, no scaling rule was available at intermediate collision velocities.

Several scaling rules were determined for multiple electron capture, but without any velocity dependence [5, 6]. Figure 5.7 presents the absolute cross sections σ_q^r for removal of exactly r electrons ($r = 1, 2$ for He, $r = 1$–3 for Ar) following Xe^{q+} + B collisions. The results of the scaling law (full curves), derived from the relation

$$\sigma_q^r = 2.710^{-13} qr/I_r^2 \sum_{j=1}^{N} \left[j(I_1/I_j)^2 \right] \tag{5.3}$$

were compared with available experiment (empty circles) and the COB model (dashed curves).

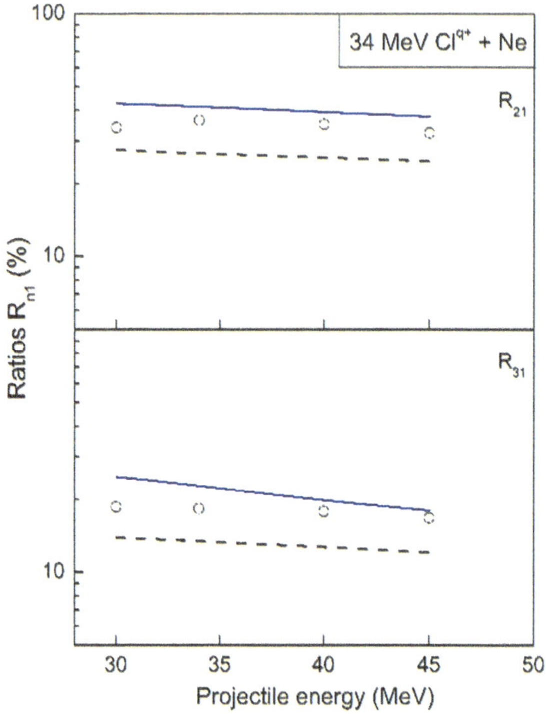

Fig. 5.6 Ratios R_{21} and R_{31} of double and triple to single capture cross section following 34 meV Cl^{12+} + Ne collisions as a function of the projectile energy. Experiment, empty circles; dashed curves, SCA calculations; full blue curves, COB model

It is clear that the scaling law is much closer to the experimental results than the COB model in the case of He target. For Ar, COB and scaling law results are very close to each other. However, the models do not account for the variations of the cross sections with q.

More recently, a great effort was paid to derive scaling laws from classical considerations for SC and SI following A^{q+} + H and A^{q+} + Li (2s, 2p) collisions, at low and high projectile energies, and then at intermediate energies [7]. Theoretical derivation of scaling laws for state-selective electron capture cross sections as a function of collision energy, initial and final bound states and projectile charge were determined. In addition, the state-selective cross sections were fitted to provide additional information about the peak locations and magnitudes.

To make predictions for electron capture cross sections and final n_f distributions, the first step was to reduce and parametrize cross sections for a few collision systems to a single and universal curve. The author used the Bohr-Lindhard model to get the capture radius at low and high projectile energies. With the knowledge of the following formula

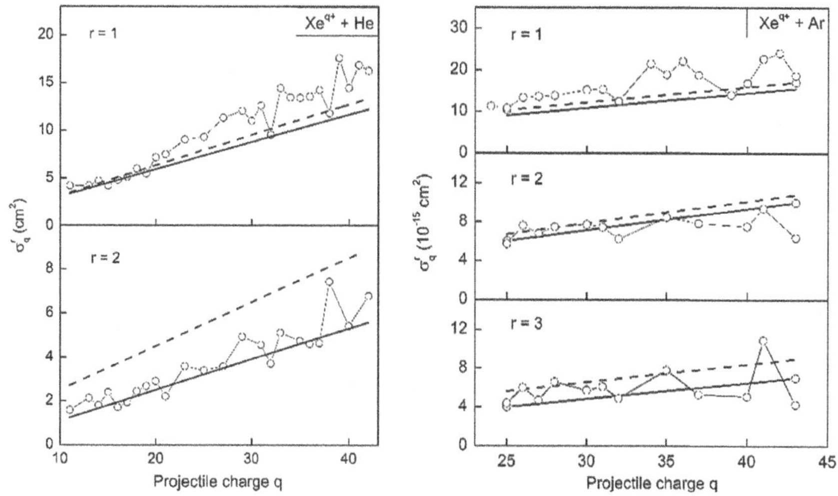

Fig. 5.7 Total cross sections for removal of exactly r electrons following Xe^{q+} + He (left side) and Xe^{q+} + Ar (right side) collisions as a function of the projectile charge q. Experiment, empty circles; COB model, dashed curves; scaling law [5], full curves

$$\sigma_{SC} = \frac{16\pi n_i^4 Z_p}{Z_t^3} \tag{5.4}$$

when $v_p \ll v_e$ and

$$\sigma_{SC} = \frac{8\pi Z_p^3}{n_i^3 Z_t^5 E_p^{7/2}} \tag{5.5}$$

when $v_p \gg v_e$

(in the above expressions, Z_p and Z_t are the projectile and target atomic numbers, respectively, and n_i is the initial principal quantum number.

The reduced cross sections $\sigma_{SC}^* = \sigma_{SC}/(qn_i^4)$ were represented as a function of the reduced projectile energy $E_p^* = E_p q^{4/7} n_i^4$ in the case of H target. For the Li target, in order to distinguish the 2 s and the 2p states, the binding energy $I = 1/2n_i^2$ was used instead of n_i.

Figure 5.8 shows σ_{SC}^* as a function of E_p^* for SC and SI following A^{q+} + Li (2s, 2p) collisions (A=He, Be, C, O and Ne). The reduced cross section follows a better universal curve than the reduced ionization cross section. At low energies, σ_{SC}^* tends towards a constant (horizontal dashed line). At high energies, σ_{SC}^* strongly decreases with E_p, since $\sigma_{SC} \propto E_p^{-7/2}$.

Then, a detailed investigation was made on partial cross sections $\sigma_{SC}(n_f)$, where n_f is the final principal quantum number on the projectile. This was done again from

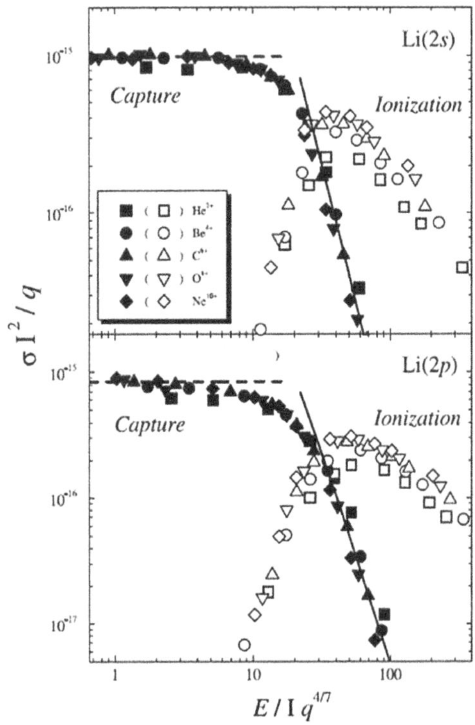

Fig. 5.8 Reduced cross sections for SC (full symbols) and SI (empty symbols) following A^{q+} + Li(2*s*, 2*p*) collisions. The dashed and full lines depict the behavior of the cross sections at low and high energies, respectively

classical considerations. The calculations by Ryufuku, which only concern H targets, were modified in order to include Z_t. The quantity $\sigma_{SC}^*(n_f) = \sigma_{SC}(n_f)I^{3/2}/q^{1/4}$ was represented as a function of the reduced quantum number $n_f I^{1/2}/q^{3/4}$ at different projectile energies (Fig. 5.9).

Each reduced cross section was fitted using two functions (full curves in Figs. 5.8 and 5.9): a Gaussian curve $f_g(n_f)$ to model the peak location n_f^o and the FWHM w of the distribution, and an exponential function $f_e(n_f)$ to reproduce the strong decreasing cross section at large values of n_f. These functions have the form

$$\begin{aligned} f_g(n_f) &= Ae^{-2\left(n_f - n_f^o\right)^2/w^2} \\ f_e(n_f) &= C\left(n_f\right)^k \end{aligned} \tag{5.6}$$

Both functions were adjusted so that $f_g(n_f) = f_e(n_f)$ at the FWHM value. The dependences of n_f^o, w and k with the reduced energy are presented in Fig. 5.10. Except for k where a universal curve can be proposed for the three possible targets, a systematic behavior of n_f^o and w cannot be deduced from the calculations. Finally, one can conclude that, despite tremendous efforts to derive scaling laws, the limitations

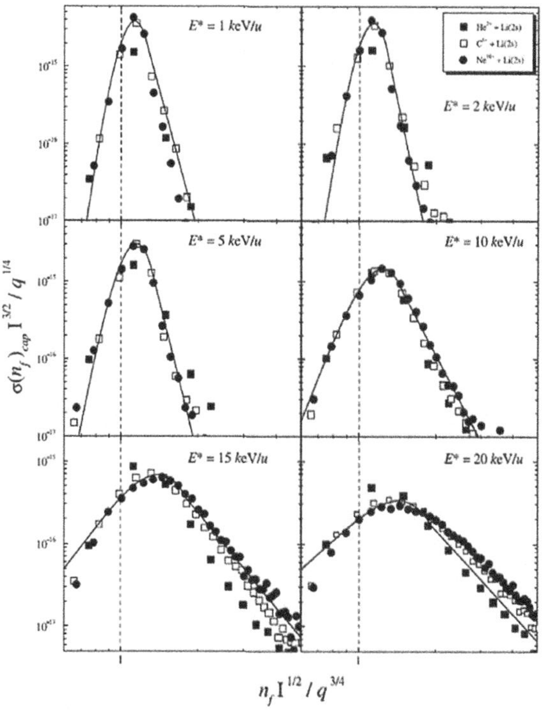

Fig. 5.9 Reduced state-selective SC cross sections as a function of the reduced final principal quantum number for different reduced projectile velocities, following A^{q+} + Li (2*s*) collisions. Each reduced cross sections was fitted with two functions (for more details see text)

are clearly seen. This work is only applied to SC, and the extrapolation to multiple capture, as well as to more complex targets, is difficult to imagine at present.

5.3 CTMC Calculations

5.3.1 Collisions Involving Alkali-Metal Series

5.3.1.1 A^{q+} + Na (n >> 3), Rb (n >> 6) Collisions

The choice of Rydberg atoms as targets, i.e. targets in which the active electron is in very highly excited sates, is motivated by the fact that electron capture will give rise to highly excited states of the projectile. Consequently, such Rydberg ions are semiclassical in nature. Therefore, the electron capture should be well described by a CTMC model. Using classical descriptions is also necessary since, for these collisions, conventional quantum mechanical methods cannot be applied in view of the large number of molecular or atomic basis to consider in the calculations. Thus, one decisive purpose of the following works was to assess the validity of the CTMC method by comparison with available experimental data.

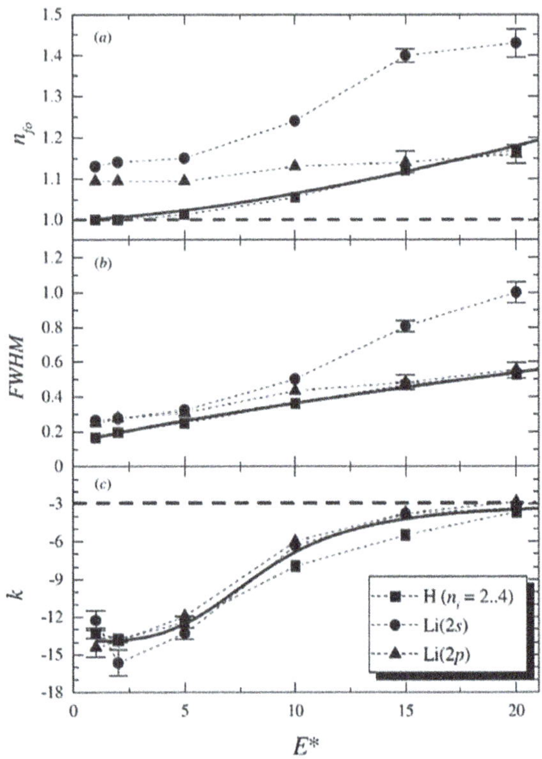

Fig. 5.10 Graphs of **a** the peak location n_f^o, **b** the FWHM value w and **c** the exponential value k as a function of the reduced collision energy E_p^*. In all three plots, the squares represent data from hydrogen targets ($n_\text{i} = 1$–4), the circles Li(2s) and the up triangles Li(2p). The full curves in (**a**) and (**b**) are the best-fit curves to the hydrogen data only. The broken curve in (**a**) represents a n_f^o value of unity. In (**c**), the full curve represents the best-fit curve using all targets. The broken line represents a value of $k = -3$, the expected asymptotic behaviour for all curves at high energies

Firstly, total capture cross sections were determined as a function of the ratio v_p/v_e (Fig. 5.11), for collisions involving Ne^+, Ar^+, Xe^+ and Na targets where the active electron was excited in the 40d state. Experiment (empty squares) was compared with

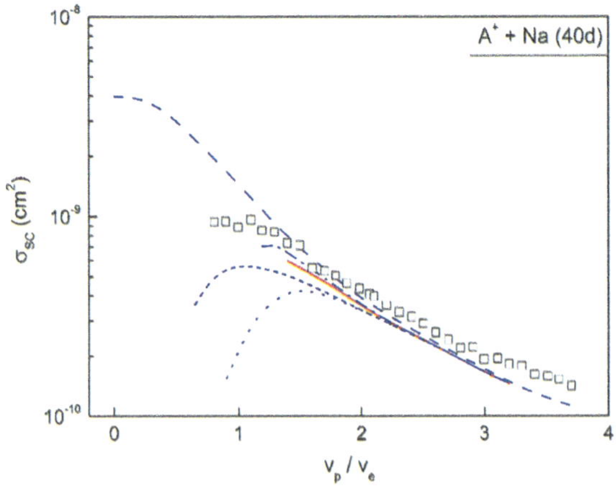

Fig. 5.11 Single capture cross section following Ne^+, Ar^+, Xe^+ + Na (28d) targets. Experiment, empty circles; quantum mechanical calculations, blue curves; CTMC calculations, red curve. For more details see [8]

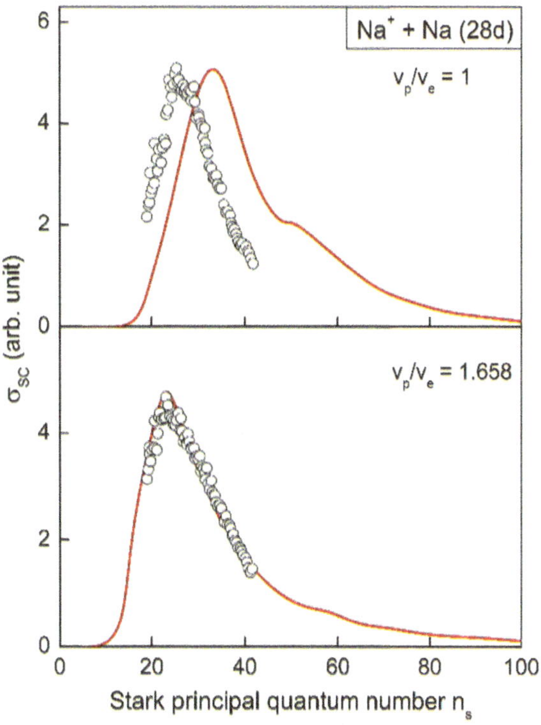

Fig. 5.12 SC cross sections as a function of the Stark principal quantum number n_s, following $Na^+ + Na$ (28d) collisions at $v_p/v_e = 1$ (top) and 1.658 (bottom) [8, 9]. Experiment, empty circles; full red curves, CTMC calculations

quantum calculations (blue curves) and CTMC calculations (red curve). Whereas all the calculations converge to the same value at large ratios, discrepancies are clearly seen at small values of v_p/v_e. The CTMC calculation was found to in good agreement with experiment. However, strong deviations would probably appear at smaller values of v_p/v_e since the CTMC method is known to be not valid in this domain.

Examples of calculated partial cross sections following SC in $Na^+ + Na$ (28d) collisions at $v_p/v_e = 1$ and 1.658 are shown in Fig. 5.12 as a function of the final principal Stark quantum number n_s, using one-electron potentials to describe the interactions between the active electron and each nucleus [8, 9]. The quantity n_s was defined as $n_s = 1/2$ for s states, $n_s = n - 1/2$ for p and d states, with n_s increasing monotonically to $n + 1/2$ for l tending towards $l_{max} = n - 1$. The calculations were compared with previous experiments. The agreement is very good for $v_p/v_e = 1.658$. For $v_p/v_e = 1$, a shift in the distribution is observed, and the calculated distribution is wider than the experimental one. In fact, as explained by the authors, by calculating the final m distributions, it was shown that final states with $m > 1$ contributed to more than 60% of the total capture cross section for $n = 33$. Consequently, the m distributions are wider than commonly assumed. The authors concluded that, due to the experimental field-ionization detection technique, the analysis in terms of adiabatic field-ionization process which assumes a preferential population o final states with $m = 0$ and 1 is not valid. In other words, the CTMC method was able to point out a failure in the experimental analysis. The agreement for 1.658 is thus probably accidental but cannot be explained.

5.3.1.2 Collisions Involving Targets in Their Quasi-Fundamental States

Total and partial cross sections

When alkali atomic targets are in their fundamental or quasi-fundamental states (for example Na (3 s) or Na (3p)), the targets can be considered in first approximation as pseudo one-electron systems. One of the benchmark systems is He^{2+} + Na (3 s), which was extensively studied experimentally and theoretically. In particular, one electron capture and ionization were analysed using the CTMC model, using one-electron model potentials [10, 11]. Figures 5.13 and 5.14 present total SC and SI cross sections, respectively, as a function of E_p. The CTMC calculations (full red

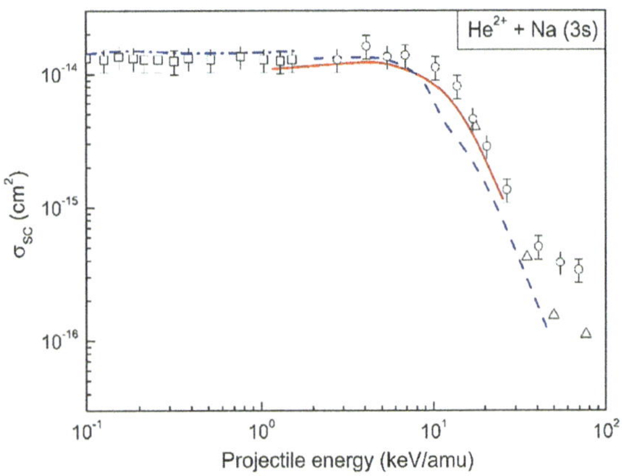

Fig. 5.13 Total SC cross sections following He^{2+} + Na (3s) collisions as a function of the projectile energy E_p. Experiment, empty squares, circles and triangles; quantum calculations, blue curves; 3B-CTMC calculations, full red curve. For more details, see [10]

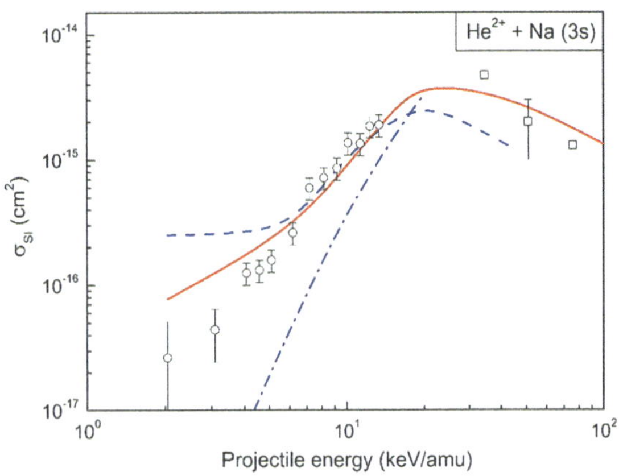

Fig. 5.14 Total SI cross sections following He^{2+} + Na (3s) collisions as a function of the projectile energy E_p. Experiment, empty squares, and circles; quantum calculations, dashed blue curve; scaling law, dotted-dashed blue curve; 3B-CTMC calculations, full red curve. For more details, see [10]

curves) were compared with available experiments (symbols) and previous quantum calculations.

The experimental SC cross sections show an almost constant behaviour below 2 keV/amu, and a rapid decrease at energies larger than 10 keV/amu, well reproduced by quantum calculations (dotted-dashed blue curve and dashed blue curve, respectively).The CTMC calculation reproduces very well the experiment in the energy range explored (1–25 keV/amu). One would expect more theoretical analysis below and above this range to check again the validity of the classical model.

For SI, the CTMC calculations (full red curve) are in very good accord with the available experiment (empty symbols). The quantum calculation (dashed blue curve) overestimates the experiment at the lowest energies. The result of a scaling law is also shown (dotted-dashed blue curve) but is found to be far from the experimental cross sections.

To get more insights on the n contributions, state selective cross sections were measured and calculated (Fig. 5.15). The CTMC results (full red curves) were compared with experiment (empty symbols) and quantum calculations (dashed blue curves) [10]. The CTMC results were in much better agreement with experiment for $n = 4$ and $n > 4$ than quantum distributions. In contrast, the CTMC calculations overestimated the low n-values. The main reason, according to the authors, resided in the fact that the classical method is expected to better reproduce high n contributions since they start to reach their classical limit.

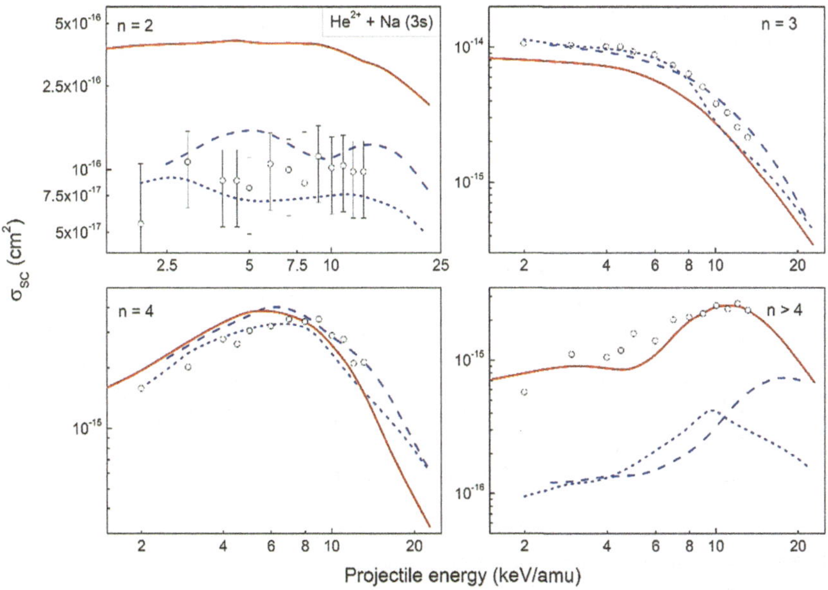

Fig. 5.15 Partial SC cross sections following He^{2+} + Na (3s) collisions as a function of the projectile energy E_p. Experiment, empty circles; quantum calculations, dashed and short dashed blue curves; 3B-CTMC calculations, full red curve [10]

Experimental and theoretical studies of partial cross sections were also performed at low energies for Ne^{8+} + Na (3s) collisions [11], in order to get evidence for oscillatory structures in the energy distributions of the n-state cross sections. As seen previously, these oscillatory structures could be interpreted classically in terms of swaps the active electron undergoes across the potential saddle before it is captured by the projectile.

The main difficulty in the CTMC method is to report the number of swaps. In the calculation, an electron swap was recorded each time the electron position vector component along the internuclear axis crossed the potential saddle position r_s, which depends onto the internuclear distance R via the relation

$$r_s = \frac{R}{\sqrt{Z_p + 1}} + aR^2 e^{-\alpha R} + bR^2 e^{-\beta R} \quad (5.7)$$

where a, b, α and β are adjustable parameters.

In Fig. 5.16 state selective capture cross sections to $n = 8$–11 are presented as a function of E_p in the range 1–10 keV/amu. The CTMC partial cross sections arising

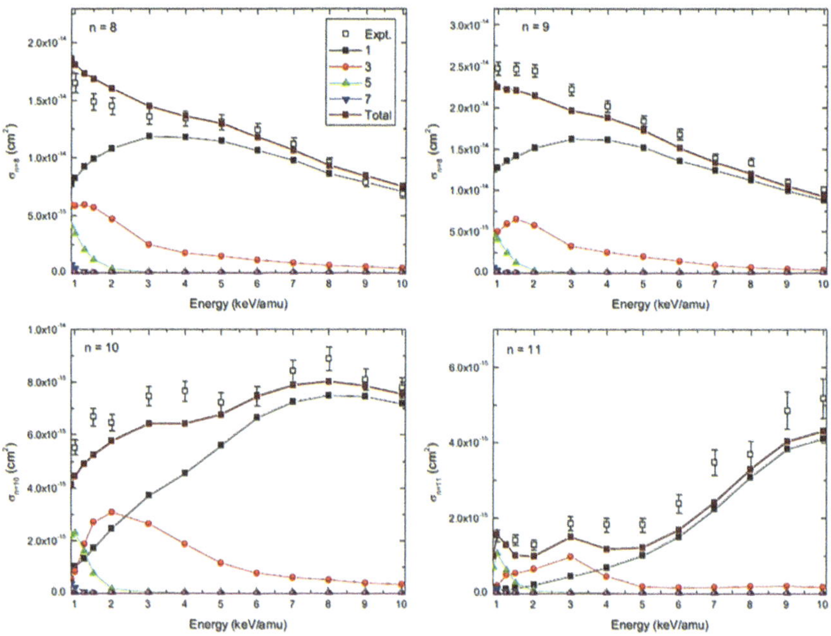

Fig. 5.16 Capture cross sections to $n = 8$–11 for Ne^{8+} + Na (3s) collisions, as a function of the projectile energy. The CTMC partial contributions from the different number of swaps are explicitly shown: 1 swap, full squares; 3 swaps, full circles, 5 swaps, full up triangles; 7 swaps, full down triangles; experiment, empty squares

from 1-, 3- and 5-swap contributions were explicitly drawn for each n value. Oscillatory structures were clearly evidenced by the experimental data (empty squares) in all cases and were reproduced by the present CTMC calculations. At the highest E_p, the 1-swap mechanism was found to dominate. It is not surprising since the active electron does not have time to make several back and forth movements before being captured. As E_p decreases, the interaction time increases so that the probability for 3 or 5 swaps also increases.

Many studies were also devoted to processes that occur in collisions involving Li targets. One advantage of such atom is the restricted number of electrons. In addition, Li atoms have many applications. As an example, neutral Li beams are used in tokamaks to provide valuable information of plasma parameters by analysing the line-radiation following collisions with other particles.

Very recently, Li^{3+} + Li (2s, $2p_0$ and $2p_1$) were studied in details theoretically in the energy range 0.1–100 keV/amu [12]. However, since no experimental data is available for this system, quantum calculations were compared with CTMC calculations performed earlier [13]. Figure 5.17 shows the results of quantum (dashed blue curves) and classical (full red curves) partial SC cross sections as a function of E_p for n varying from 3 to 7. The agreement between both theories is remarkable, except for $n = 3$ and $n = 5$, for which strong differences occur at energies smaller than 10 keV/amu and 1 keV/amu, respectively. It is again clear that experimental data are missing to judge the validity of one or the other model.

Differential cross sections

The determination of differential cross sections is a challenge, as it was shown in the previous chapters. Of course, the challenge is all the greater as the collision system becomes more complex. On the other hand, since the approximation of a three-body system, including two nuclei and an active electron, remains valid in the case of alkali targets, the challenge remains quite easy to take up.

As for the case of H and He targets, many experiments and calculations were focused in SDCS and DDCS, in order to get more insights on the collision dynamics. For example, experimental transverse momentum distributions were extracted from He^{2+} + Na (3 s) collisions and compared with available 3B-CTMC calculations [10] (Fig. 5.18). The DDCS were normalized to a peak value of 1, and the transverse momenta were scaled with the square root of E_p. Therefore, in this representation, the abscissa was approximately inversely proportional to the impact parameter. The Fig. 5.18 compares the experiment DDCS (dashed blue curves) for capture into $n = 4$ at E_p of 2, 3, 5 and 10 keV/amu with calculations (full red curves). Despite the relatively good agreement between experiment and theory, the calculated distributions were found to be systematically much broader than the experimental data. In addition, the experimental data exhibited structures that are not present in the calculations. Finally, the experimental and calculated positions of the maxima were found to be similar, except at the highest energy. So far, no explanation was given to explain these discrepancies.

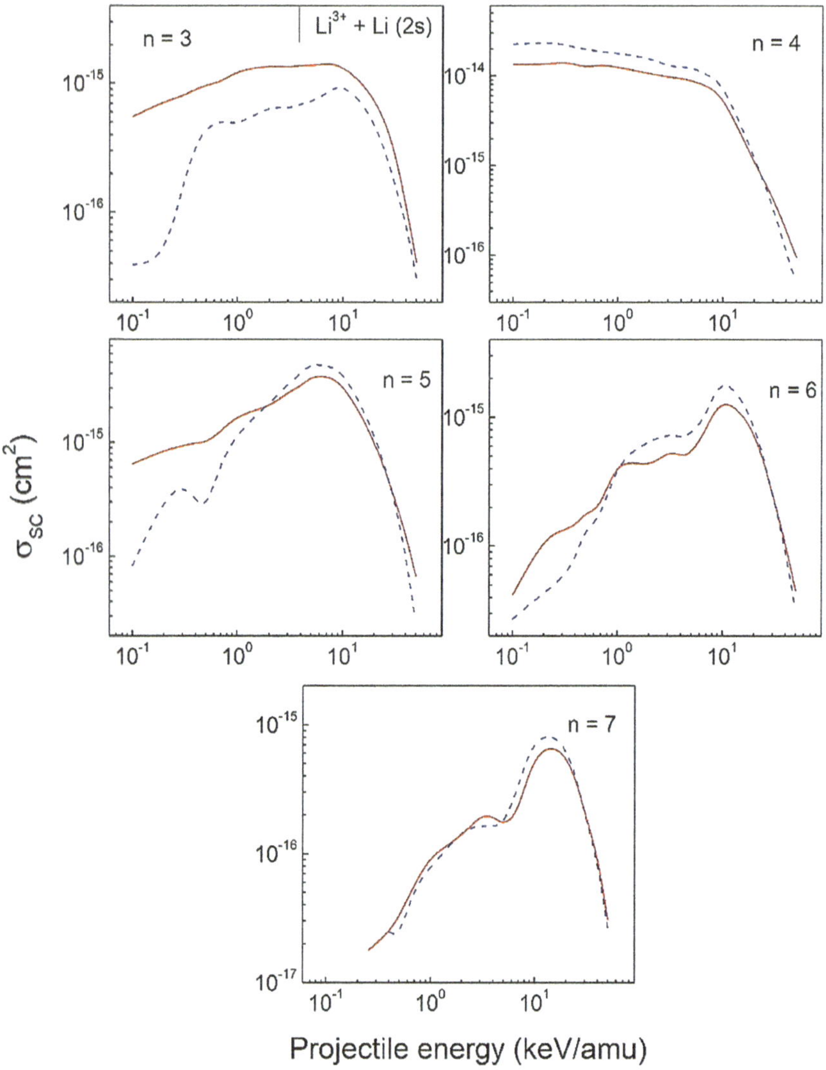

Fig. 5.17 Partial SC cross sections following Li $^{3+}$ + Li (2s) collisions as a function of the projectile energy, for *n* ranging from 3 to 7. Quantum calculations, dashed blue curves; CTMC calculations, full red curves

5.3.2 Collisions with Noble Targets

When highly charged ions collide with rare gases (Ne, Ar, Xe, Kr), cross sections for multiple capture (MC) at low energies and multiple ionization (MI) at high energies are found to be large, since these targets have a huge amount of available outer-shell electrons. As an example, for the collision Ar^{18+} + Ar at E_p = 1 keV/amu, the

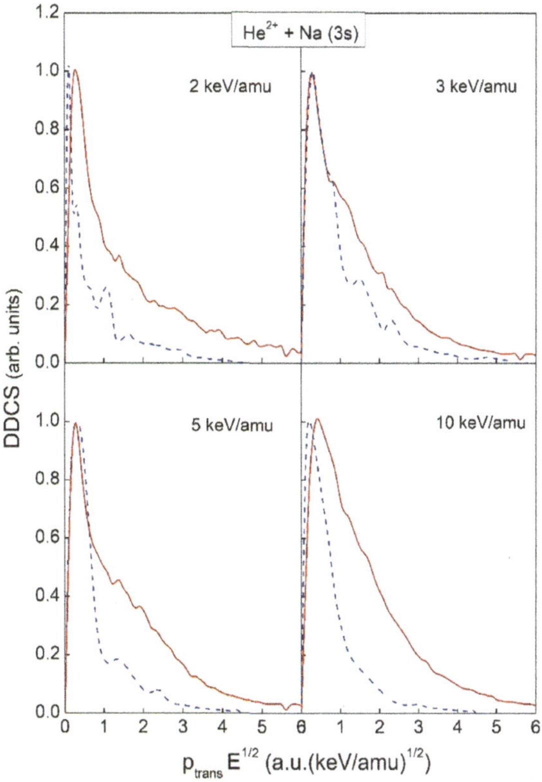

Fig. 5.18 DDCS following He $^{2+}$ + Na (3s) collisions at projectile energies of 2, 3, 5 and 10 keV/amu, as a function of the transverse momentum scaled with the square root of E_p. Experiment, dashed blue curves; 3B-CTMC calculations, red full curves [10]

cross sections for removing 4 up to 10 electrons are quite constant and of the order of 10^{-16}–10^{-15} cm^2 [14]. The advantage of the CTMC method is that, in the first approximation, the active electrons may be considered as independent, even though they have the same binding energy. This approximation is valid and thus useful when calculating total cross sections. However, it is clear that, for such targets, electron correlation will play an important role.

5.3.2.1 Total and Partial Cross Sections for Target or Projectile Ionization

In the case of multiple electron targets, only a few data are available on multiple capture or multiple ionization. Experiments and calculations were mainly focused onto SC and SI, or on total electron loss, even though some data exist concerning the removing of exactly n electrons.

In this section, we will concentrate on the dependence of total cross sections with the projectile energy E_p and the projectile charge q. Target ionization was mainly studied at intermediate and high energies. At such energies, many experiments

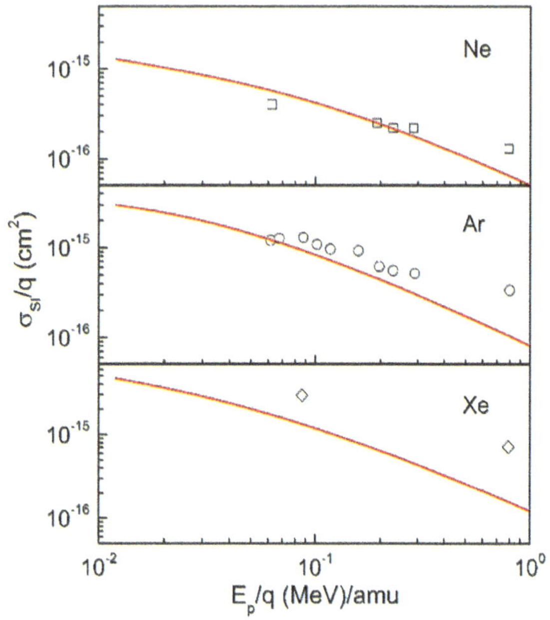

Fig. 5.19 SI reduced total cross sections following A^{q+} + Ne, Ar and Xe targets ($q = 4 - 54$), as a function of the reduced energy E_p/q. Experiment, Ne, empty squares, Ar, empty circles and Xe, empty diamonds. 3B-CTMC calculations, full red curves [15]

were devoted to the systematic determination of total ionization cross sections σ_I. Figure 5.19 summarizes scaled cross sections σ_I/q as a function of the reduced energy E_p/q for projectile with charges varying from 4 to 54, and for Ne (open squares), Ar (open circles) and Xe targets (open diamonds) [15]. The experimental data were compared with 3B-CTMC calculations (full red curves).

The results were found to be reasonable for Ar and Ne, except for large values of E_p/q. For Xe, the agreement was found to be poor, since the calculations underestimate the experiment by a factor of more than 2. At this stage, unfortunately, no discussion was began to explain the observed discrepancies.

More recently, projectile ionization, also called electron loss[1] by the authors [16] was investigated theoretically at low and intermediate energies for singly charged ions colliding on noble gases. The results from energy-deposition (ED) model[2] were compared with available data and the recommended results. Some results derived from CTMC calculations were also represented.

Two typical examples are given in Fig. 5.20 for A^+ + Ne and Ar as a function of E_p. Experimental results (empty squares and circles) were mainly available below 100 keV/amu. For A^+ + Ne collisions, neither the recommended data (dashed blue curve) or the ED model were able to fit the experiment, in contrast with A^+ + Ar at low energies. At high energies, the CTMC calculations (full red curves) are in good agreement with experiment for A^+ + Ar, whereas they deviate for A^+ + Ne

[1] The electron loss process definition depends onto the authors. Here, it characterizes the ionization of the projectile, while in most cases it means charge changing.

[2] The ED model is not detailed here for reasons of clarity. For more details see [16].

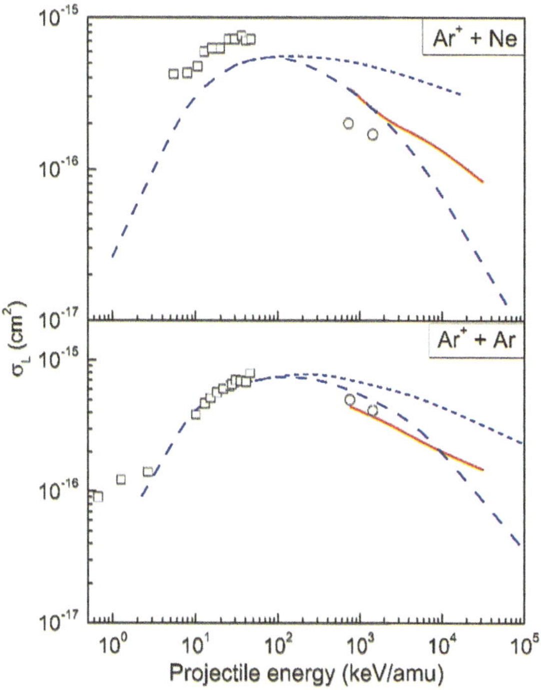

Fig. 5.20 Electron loss cross sections following $A^+ + Ne$ and $A^+ + Ar$ collisions as a function of E_p. Experiment, empty squares and circles; ED model, short dashed blue curves; recommended data, dashed blue curves; CTMC calculations, full red curves

by a factor of about 2. It is clear that classical calculations would be necessary to perform in a larger range of energies, even though the domain of validity of CTMC is exceeded.

Multiple ionization was also studied in more details by determining cross sections for ionizing exactly n electrons. This studies could be performed because of the development time-of-flight techniques in the early 1980's. One of the questions was the possibility to consider active electrons as independent, leading to binomial statistics with respect to the degree of multiple ionization, or have electron correlation effects a measurable influence on the partial cross section σ_n?

Two results for σ_n, with $n = 3$ (empty squares) and 10 (empty circles) as a function of E_p are shown in Fig. 5.21 in the case of $U^{q+} + Ar$ [17]. The experimental results were compared with the multi-body nCTMC model (full red curves), where K-, L- and M-shell electrons were explicitly included.

Auger effects were taken into account by calculating the total electron energy of the residual Ar ion after the collision. The agreement between experiment and calculation is excellent, showing that dynamic electron correlation, i.e. electron correlation during the collision, is not necessary to include. However, electron–electron interaction plays a decisive role, via the Auger effect, as seen in Fig. 5.22 for 120 meV/amu U^{q+} impact. This is due to the fact that not only M-shell electrons are involved during the collision, but also L-shell electrons. In addition, valence electrons may

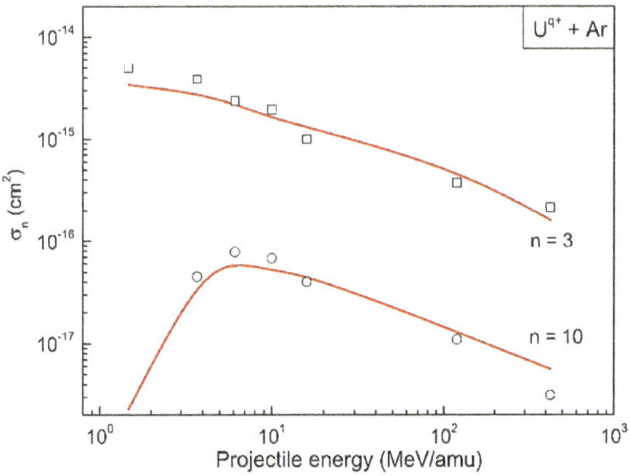

Fig. 5.21 Partial ionization cross sections for removing n = 3 and 10 Ar electrons following U^{q+} impact as a function of the projectile energy. Experiments, empty squares and circles; nCTMC calculations, full red curves

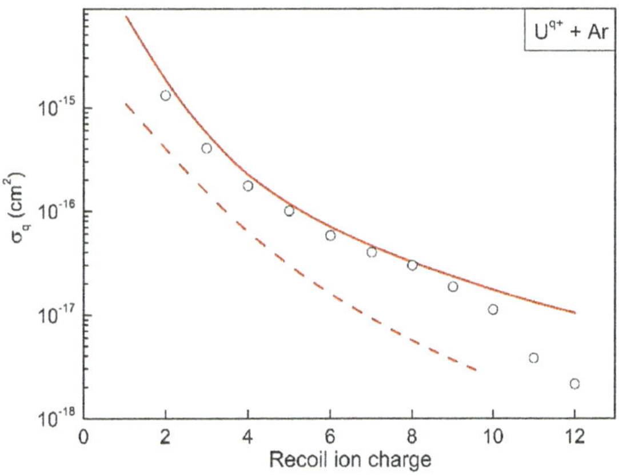

Fig. 5.22 Recoil ion charge state distribution following multiple ionization in 120 meV/amu U^{q+} + Ar collisions [17]. Experiment, empty circles; nCTMC calculations without autoionization, dashed red curve; nCTMC calculation after inclusion of autoionization, full red curve

also be excited, giving rise to multiply excited states that decay via the autoionization process. When autoionization is not included (dashed red curve), the calculations underestimate the experiment by a factor of at least 2. After the inclusion of the Auger effect, the theoretical result fits very well the experimental cross sections, except for recoil ion charges larger than 10.

It is clear from the examples given above that target ionization is the main process of electron emission that was studied. However, it is well known that the projectile itself may also be ionized. One of the challenges was thus to develop a theory that takes into account electron emission from the target and the projectile. Such process, that provides a unique opportunity to study the interplay between electron correlation and screened nuclear-electron interaction, was examined in details for the collision system He^+ + Ne at projectile energies varying from 1 meV up to 4 meV [18]. In this study, the nCTMC method was utilized. The electron correlation between the Ne electrons was neglected, whereas the interaction between the He electron and each L-shell Ne electron was included.

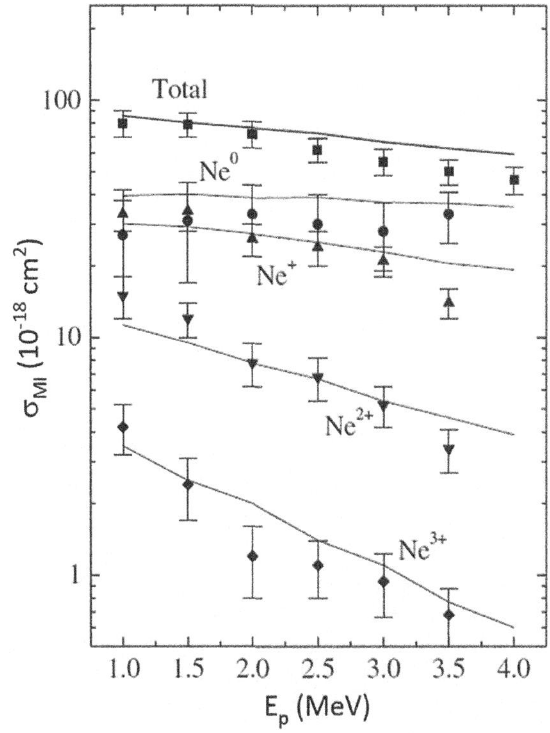

Fig. 5.23 Total cross for simultaneous ionization of the target and the projectile following the reaction $He^+ + Ne \rightarrow He^+ + Ne^{i+} + (i + 1)\ e^-$ ($i = 0$–3) as a function of E_p. Experiment, full objects; nCTMC calculations, full curves

The cross sections for the reaction $He^+ + Ne \rightarrow He^+ + Ne^{i+} + (i + 1)\ e^-$ ($i = 0$–3) are displayed in Fig. 5.23 as a function of E_p. The agreement between the experiment and the calculations was found to be remarkable for all the degrees of ionization. More interesting, this method could predict that the observation of the relative azimuthal angle correlation between the ionized projectile and the target electrons would provide a useful to probe two-centre electron correlation dynamics.

5.3.2.2 Total and Partial Cross Sections for Electron Capture

Electron capture was investigated for singly and multiply charged ion impacting on noble gases, at low and high projectile energies. Again, the use of the CTMC approach was motivated by the fact that a quantum mechanical treatment of collisions involving more than two electrons is prohibitively difficult due to the large number of channels involved. It was shown that multiple capture (MEC) had to be included in the calculations, since it represents a large percentage of the total electron capture. For example, in the case of 4.55 keV/amu Ne^{10+} + Ne and Ar, the MEC process represents ~46% and 54% of the total capture [19]. Unfortunately, even though the order of magnitude is good, the CTMC calculations based on the

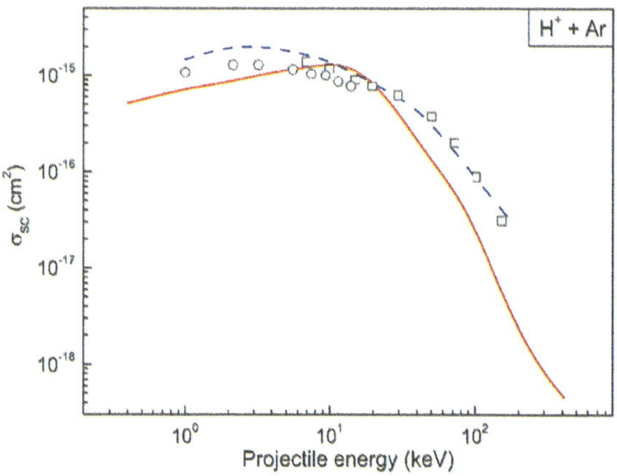

Fig. 5.24 SC cross sections following H^+ + Ar collisions as a function of E_p [19]. Experiments, empty circles and squares; quantum calculations, dashed blue curve; 3B-CTMC calculations, full red curve

standard microcanonical distributions underestimated these fractions (33% and 38%, respectively). The difference between experiment and calculation may have several reasons. First, electron correlation, which plays a decisive role in such low energetic collision systems, is not taken into account in the calculations. Moreover, the Auger effect may induce effects that would change the percentages.

In this section, two detailed examples are given, SC at low energies, and MC at high energies. At low energies, the collision system H^+ + Ar was revisited a few years ago [20]. The goal was to provide preliminary results on DC, that gives rise to H^- ions. This study followed previous experiments involving OH^+ ions impacting on Ar at low energies. Since H^- is very difficult to study classically, DC cross sections were deduced from SC cross sections using the CTMC method with uncorrelated *Ar* electrons. To validate this method, SC cross sections were calculated in a large range of projectile energies (500 eV–400 keV) and compared with previous experiments and quantum calculations (Fig. 5.24).

The CTMC calculations (full red curve) were found to agree well with the experiment (empty circles and squares) for E_p smaller than 30 keV, but under estimated the data at larger values of E_p by a factor of approximately 2, whereas the quantum calculations fitted very well the data. Despite some discrepancies, the relatively good agreement shown by the classical calculations was enough to be confident with the method employed and to go further by determining the DC cross sections (full red curve Fig. 5.25). Unfortunately, the quantitative agreement with the experiment was poor, qualitatively and quantitatively. DC cross sections using the CTMC model were five times larger than the experimental cross sections (empty circles, squares and diamonds) between 5 and 40 keV. At smaller E_p, the CTMC values increase, reaching 10^{-16} cm^2 at 400 eV. In contrary, the previous quantum calculations were smaller than experimental data by a factor of ~4 in the whole range of energies. We could only note a qualitative agreement in the shape of the cross sections variation, since a maximum and a minimum were observed at ~20 keV and 10 keV, respectively, for the three cases. The disagreement in the absolute values is certainly due to the fact that the *Ar* 3 l electrons are treated independently. If electron correlation could have

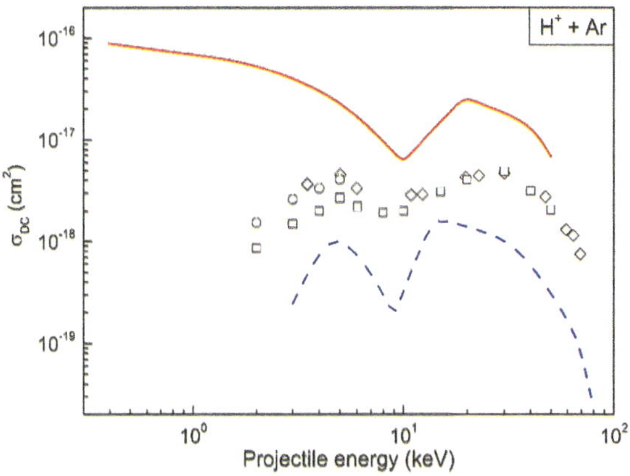

Fig. 5.25 DC cross sections following $H^+ + Ar$ collisions as a function of E_p [19]. Experiments, empty circles, squares and diamonds; quantum calculations, dashed blue curve; 3B-CTMC calculations, full red curve

been introduced in the classical calculation, *Ar* excitation and Auger effect would have play a non negligible role and would have contributed to a decrease of the cross sections. In addition, the introduction of electron correlation would have induced the production of excited states of the projectile, which is unstable, increasing the diminution of the cross sections.

As mentioned above, electron capture was also investigated at high energies. For example, $C_n^+ + Ar$ collisions were studies experimentally and theoretically at energies of 125 keV/amu, corresponding to a velocity of 2.25 a.u. [21]. The experimental SC and DC cross sections were compared with predictions from the independent atom and electron (IAE) model developed to describe cluster-atom collisions. The probabilities required for the IAE model were determined by CTMC calculations. In the calculations, pure SC process and DC followed by projectile ionization were included to describe the final process. The results for SC and DC cross sections as a function of the number *n* of *C* atoms involved are shown in Fig. 5.26. In the case of SC, the calculations (full red curve) agree rather well with the experiment

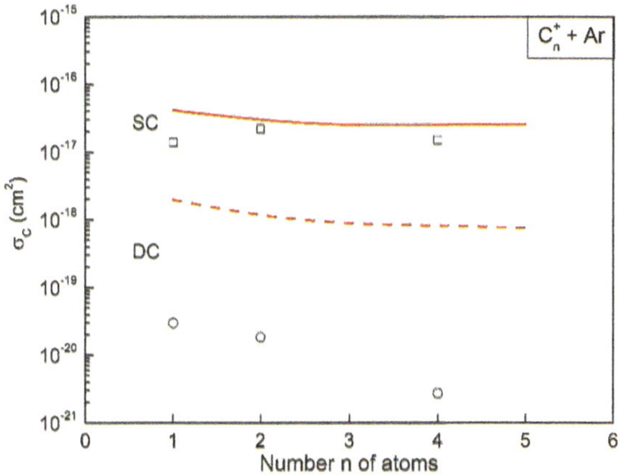

Fig. 5.26 SC and DC following $C_n^+ + Ar$ collisions at a projectile velocity of 2.25 a.u., as a function of the number *n* of C atoms involved. Experiment, empty squares and circles for SC and DC, respectively; CTMC calculations, full and dashed red curves for SC and DC, respectively [21]

90° (right side of Fig. 5.29). This result was also observed in the case of U^{32+} + Ne collisions.

5.3.2.4 SDCS and MDCS in Capture Processes

The study of electron capture from multielectron capture targets has many applications. This process is important in the understanding of the collision itself between highly charged ions coming from the solar wind and the coma. As an example, X-ray emission originating from charge-exchange collisions was found to be responsible for the cometary X-ray emission discovered by the ROSAT satellite in 1996. From this discovery, many studies were performed experimentally and theoretically to reproduce the astrophysical observations. Since the range of impact energies, the large projectile charges involved and the multielectronic nature of the targets make the implementation of quantum mechanical models prohibitive, the theoretical works in this area were mainly performed using classical mechanics based on different versions of the CTMC model [14, 23].

The goal of such studies was first to calculate emission cross sections after SC and MC, as well as probabilities of $nl \rightarrow n'l'$ transitions (nl and $n'l'$ are the quantum numbers that characterize the initial and final states of the projectile, respectively), that require the knowledge of l-distributions, and then to reproduce the experimental spectra, with the knowledge of the relative abundance of the ions and the width of the detectors. The initial populations of nl states after capture were calculated by means of the CTMC method, while emission cross sections were determined with branching ratios for the relevant transitions.

To validate the method, Ne and Ar targets were extensively used. Different models were developed with several model potentials (hydrogenic potentials (1), central potentials (2), classical potentials reproducing quantum momentum distributions (3), Clementi expansions (4)). Figure 5.30 presents an example of line emission distributions for Ar^{18+} + Ar collisions at four projectile energies [14]. The experimental cross sections (thick lines) were compared with CTMC calculations using (3) and (4) potential expansions. Excellent agreement was observed and the shoulder on the Lyman-α peak, located at ~3100 eV, which is a signature of multiple electron capture, was clearly reproduced in the multiple electron calculations (model (4)).

A few years later, the role of multiple electron capture was evidenced in A^{q+} (q = 10, 17, 18, 36) + Ar collisions by means of the CTMC method, using an eight-electron scheme to theoretically study photonic spectra following capture [23]. The previous model was extended by accounting in detail the energy deposition needed for many-electron removal, and by using a Auger decay scheme developed previously for Ar^{q+} + Ar collisions [24]. Because of the intrinsic complexity of Auger decay, many approximations were made, that are not given here (for more details, see [24]).

A line emission spectrum following 4 keV/amu Ne^{10+} + Ar is presented in Fig. 5.31 [23]. The experimental results (open circles), obtained at NIST by using an EBIS trap in the extraction mode of operation, were normalized to the present CTMC calculations. To match the experimental cross sections, the calculations were

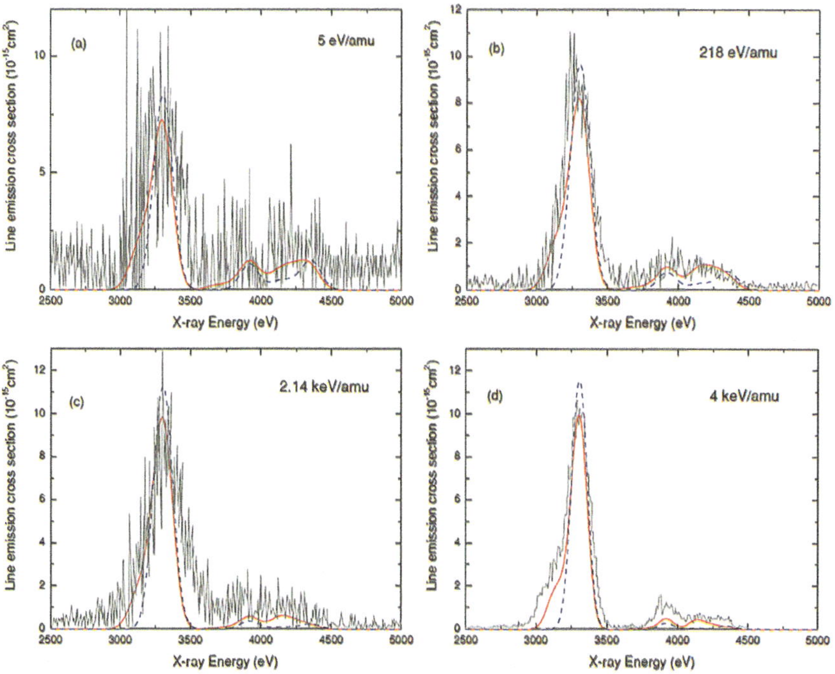

Fig. 5.30 Line emission cross sections following MC in Ar^{18+} + Ar collisions at four projectile energies [14]. Experiment, thick black lines; CTMC calculations using model (3) and (4), dashed blue curves and full red curves, respectively

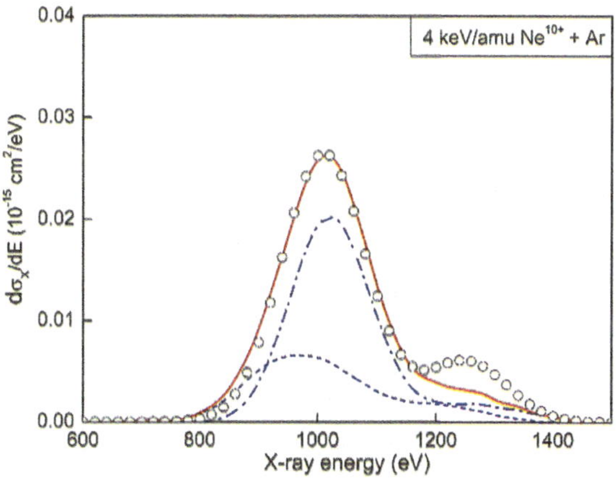

Fig. 5.31 Line emission cross sections following MC in Ne^{10+} + Ar collisions. Experiment, empty circles; SC contribution, dotted-dashed curve; ADC contribution, dashed curve; NADC contribution, short dashed curve; sum, red curve [23]

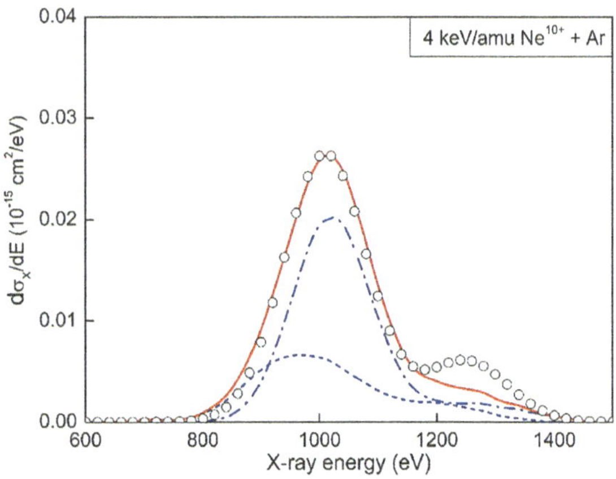

Fig. 5.32 Line emission cross sections following MC in Ne^{10+} + Ar collisions. Experiment, empty circles; SC contribution, dotted-dashed curve; NADC contribution, short dashed curve; sum, red curve

convoluted by means of FWHM Gaussian curves. The different dashed curves show the separate contributions originating from SC (dotted-dashed curve), autoionizing DC (ADC, dashed curve) and non autoionizing DC (NADC, short dashed curve). The sum of all the contributions is depicted by the full red curve.

The question arises whether the Auger contributions is well described and necessary to include in the calculations to fit the experiment, since they were found to be relatively small, compared to SC or NADC processes. For this reason, the same spectra are represented in Fig. 5.32, after removal of the Auger contribution. The result is very similar to that obtained when the Auger contribution is included, indicating that the calculations of the Auger part has to be questioned.

5.3.3 Combination of Processes

The CTMC method is a unique opportunity not only in the determination of cross sections for single processes, but also in the analysis of more particular processes, such as K-shell vacancy production, transfer ionization (TI), or Fermi-shuttle ionization. To illustrate the ability of the CTMC to give some answers, one example for each mentioned process is given below.

Twenty years ago, single and double K-shell vacancy production was analysed both experimentally and theoretically for N^{7+} + Ti collisions, in the collision energy range 4–26 meV/amu [25]. Experimental and theoretical studies of such process are very important for mapping the limits of the applicability of the independent electron approximation (IEA) which is usually considered to be valid for the inner shell vacancy production process. Consequently, attention was focused on SI and DI, and finally on the ratio $R = \sigma_{DI}/\sigma_{SI}$. Two classical models were used. First, IEA was applied by neglecting the electron correlation before, during, and after the collision. In the second model, a sudden change in the calculation was operated after one of the

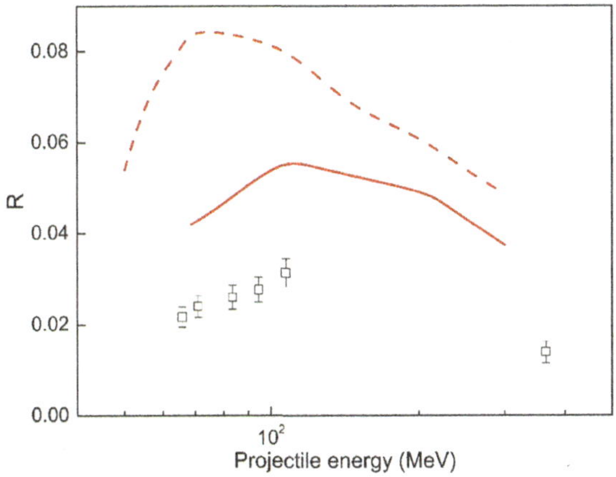

Fig. 5.33 Ratio R of Double to single K-shell vacancy following N^{7+} + Ti collisions, as a function of the collision energy [25]. Experiment, empty squares; CTMC calculations, dashed red curve; CTMCF calculations, full red curve

electrons was captured by the projectile. The captured electron was frozen, inducing a change of the effective charge of the projectile. Thus, a repulsive force between the two electrons was roughly modelled. These frozen-electron CTMC calculations were referred to as CTMCF.

The ratio R calculated by means of both classical model is presented in Fig. 5.33 as a function of E_p, and compared with experiment (empty squares).

The CTMC ratio is clearly overestimated. In contrast, the introduction of screening in the projectile charge induces a strong diminution of R, showing, according to the authors, that the electron correlation in the final state may play a role.

It is well known that emitted electrons from a target are able to be captured into continuum sates of the projectile. An example of such process has been given in Chap. 4, for the collisions system Au^{53+} + He. When a second electron is involved, it recoils with the speed of the projectile, resulting in a pronounced peak located around 90° in the angular distribution of the emitted electrons with the matching velocity $v_e = v_p$ (cusp electrons), indicating a strong correlation between the electrons. In 2001, a full analysis of electron capture to the continuum (ECC), which consists of a simultaneous transfer of an electron to the continuum and the ionization of a second electron, was performed [26]. Associated cross sections were compared with TI which is, in the present work, a process that combines capture into the continuum and a capture into bound states of the projectile. To avoid difficulties due to the large number of active electrons, the He target was favoured. Nevertheless, the Ar target was also used for comparison. The experimental data were interpreted with CTMC and CDW-EIS calculations. According to the authors, this work was the first attempt to give a quantitative description of such particular process.

The ratio σ_{TI}/σ_{ECC} is presented in Fig. 5.34 as a function of E_p, and restricted to He^{2+} + Ar collisions.

The experimental ratio was found to decrease exponentially when increasing E_p. This tendency is confirmed by the CTMC calculations. The theoretical ratio fitted rather well the experiment at $E_p < 100$ keV and were overestimated above, by a

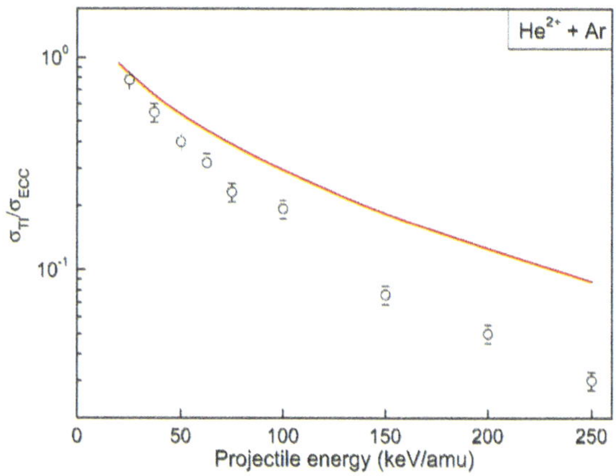

Fig. 5.34 Ratio σ_{TI}/σ_{ECC} as a function of E_p for $He^{2+} + Ar$ collisions. Experiment, empty circles; CTMC calculations, full red curve [26]

factor of 5 at the largest energy. This ratio is an interesting quantity since it is a measure of the strength of the two-electron over the one-electron process. In other words, this ratio is sensitive to electron correlation effects.

It has to be noted that only the 3B-CTMC model was used, so that electron correlation was completely neglected. The introduction of the electron electron interaction is known to be feasible for a He target, but is difficult to implement in the case of Ar.

The third example concerns CTMC calculations performed for the collision system $N^+ + Ar$ at $E_p = 15$ keV to determine the ionization cross sections as a function of the emission angle and energy of the ejected electrons, in order to understand how the electrons gain energy during the multiple scattering with the projectile and the target nuclei [27]. Figure 5.35 shows the measured electron spectrum emitted at 135° (empty circles and full black line).

In addition to pure ionization, contributions of Auger decay following multiple excitation of the target were clearly visible at energies smaller than 20 eV and larger

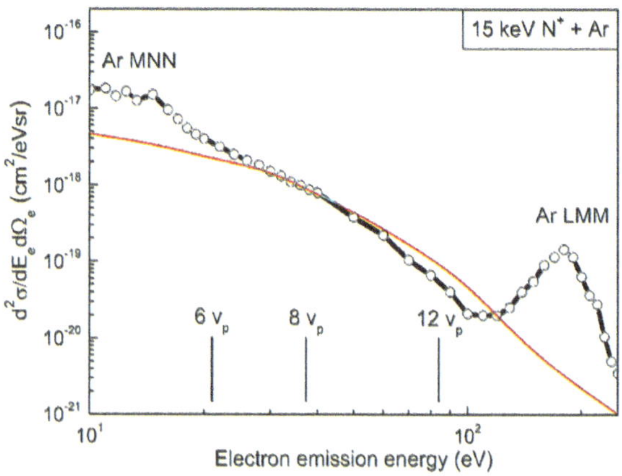

Fig. 5.35 Doubly differential cross section following 15 keV $N^+ + Ar$ collisions as a function of the emitted electron energy [27]. The two structures in the experimental spectrum (empty circles) originate from Auger deexcitation of Ar. CTMC calculations, full red curve

than 100 eV. The result was compared with CTMC calculations (full red curve) in absolute scale. Most of the CTMC results, above 10 eV, felt into the multiple scattering category (note the multiplying factors 6, 8 and 12 of v_p), indicating that the acceleration of the emitted electron during the collision could dominate ionization.

References

1. Niehaus, A.: A classical model for multiple-electron capture in slow collisions of highly charged ions with atoms. J. Phys. B **29**, 2925 (1986)
2. Ito, H., Chihara, Y., Suzuki, Y., Hirayama, T., Koizumi, T.: Multiple charge transfer by slow multi-charged Xe ions. J. Phys. Conf. Ser. **58**, 069 (2007)
3. Ishii, K., Okuno, K.: Modified classical over barrier model for multi-charged ions in collisions with multiple electron targets at energies below 1 keV/u. Plasma Sci. Techno. **12**, 369 (2010)
4. Man, Z., Zhenye, W., Sufen, L., Peng, X., Haoxin, Z.: Multiple ionization cross sections of Ne and CO induced by very high-q fast projectiles (q/v > 1). J. Appl. Math. Phys. **6**, 2343 (2018)
5. Selberg, N., Biedermann, C., Cederquist, H.: Semiempirical scaling laws for electron capture at low energies. Phys. Rev. A **54**, 4127 (1996)
6. Selberg, N., Biedermann, C., Cederquist, H.: Absolute charge exchange cross sections for the interaction between slow Xe^{q+} ($15 \leq q \leq 43$) projectiles and neutral He, Ar and Xe. Phys. Rev. A **56**, 4623 (1997)
7. Cornelius, K.R., Wojtkowski, K., Olson, R.E.: State-selective cross sections scalings for electron capture collisions. J. Phys. B **33**, 2017 (2000)
8. Pascale, J., Olson, R.E., Reinhold, C.O.: State-selective capture in collisions between ions and ground- and excited-state alkali-metal atoms. Phys. Rev. A **42**, 5305 (1990)
9. Mac Adams, K.B., Martin, N.L.S., Smith, D.B., Rolfes, R. G., Richards, D.: Electron loss from Na Rydberg atoms by ion impact. Phys. Rev. A **34**, 4661 (1986)
10. Knoop, S., Olson, R.E., Ott, H., Hasan, V.G., Morgenstern, R., Hoekstra, R.: Single ionization and electron capture in He^{2+} + Na collisions. J. Phys. B **38**, 1987 (2005)
11. Otranto, S., Blank, I., Olson, R.E., Hoekstra, R.: Oscillatory pattern in angular differential ion-atom charge exchange cross sections: the role of electron saddle swaps. AIP Conf. Proc. **27**, 1525 (2013)
12. Ma, M.X., Kon, B.H., Liu, L., Wu, Y., Wang, J.G.: Electron capture of Li^{3+}ions with ground and excited states of Li atoms. Chin. Phys. B **29** 013401 (2020)
13. Otranto, S., Hoekstra, R., Olson, R.E.: Role of electron saddle swaps in the photon spectra following Li^{3+} charge exchange with H* (n = 2=, Na (3s), Na* (3p) and Li (2s) targets. Phys. Rev. A **89**, 022705 (2014)
14. Otranto, S., Olson, R.E.: X-ray emission cross sections following Ar^{18+} charge-exchange collisions on neutral argon: the role of the multiple electron capture. Phys. Rev. A **83**, 032710 (2011)
15. Schlachter, A.S.: Collisions of highly stripped ions at MeV energies in gas targets: Charge transfer and ionization. IEEE Trans. Nucl. Sci. **28**, 1039 (1981)
16. Shevelko, V.P., Kato, D., Litsarev, M.S., Tawara, H.: The energy-deposition model: Electron loss of heavy ions in collisions with neutral atoms at low and intermediate energies. NIFS Data **112**, 1 (2010)
17. Berg, H., Dörner, R., Kelbch, C., Kelbch, S., Ullrich, J., Hagmann, S., Richard, P., Schmidt-Böcking, H., Schlachter, A.S., Prior, M., Crawford, H.J., Engelage, J.M., Flores, I., Loyd, D.H., Pedersen, J., Olson, R.E.: Multiple ionization of rare gases by high-energy uranium ions. J. Phys. B **21**, 3929 (1988)
18. Fiol, J., Olson, R.E., Santos, A.C.F., Sigaud, G.M., Montenegro, E.C.: Simultaneous projectile and target ionization in He^{+} + Ne collisions. J. Phys. B **34**, L503 (2001)

19. Ali, R., Neill, P.A., Beiersdorfer, P., Harris, C.L., Rakovich, M.J., Wang, J.G.. Schultz, D.R., Stancil, P.C.: On the significance of the contribution of multiple-electron capture processes to cometary X-ray emission. Astrophys. J. **629**, L125 (2005)
20. Frémont, F.: Electron capture and single ionization in $H^+ + Ar$ collisions: classical calculations. J. Phys. B **49**, 065206 (2016)
21. Labaigt, G., Jorge, A., Illescas, C., Bérof, K., Dubois, A., Pons, B., Chabot, M.: Electron capture and ionization processes in high velocity , C + Ar and , C + He collisions. J. Phys. B. **48**, 075201 (2015)
22. Ullrich, J., Olson, R.E., Schmidt-Böcking, H., Schmidt, S., Dörner, R., Dangendorf, V., Berg, H.: Multiple ionization and collective electron emission in MeV/u uranium-ion rare gas collisions. J. de Phys. Colloques **50**, 29 (1989)
23. Otranto, S., Cariatore, N.D., Olson, R.E.: X-ray emission produced in charge-exchange collisions between highly charged ions and argon: Role of the multiple electron capture. Phys. Rev. A **90**, 062708 (2014)
24. Ali, R., Cocke, C.L., Raphaelian, M.L.A., Stockli, M.: Multielectron processes in 10 keV/u $Ar^{q+}(5 \leq q \leq 17)$ collisions. Phys. Rev. A **49**, 3586 (1994)
25. Sulik, B., Tokési, K, Awaya, Y., Kambara, T.. Kanai Y.: Single and double K-shell vacancy production in N^{7+}+ Ti collisions. Nucl. Instr. Meth. Phys. Res. B **154**, 286 (1999)
26. Sarkadi, L., Lugosi, L., Tokési, K., Gulyas, Kövér, A.: Study of the transfer ionization process by observing the electron cusp in 100–300 keV He^{2+}+ He collisions. J. Phys. B **34**, 4901 (2001)
27. Tokési, K., Tskhakaya, D., Coster, D.: Atomic data for integrated modelling-Fermi-shuttle type ionization as a possible source of high energy electrons. EPJ Web Conf. **79**, 02003 (2014)

Chapter 6
Collisions $A^{q+} + H_2$

Abstract Since the CTMC model was successful enough to reproduce at least qualitatively many experiments, even in the details, in the case of atomic targets, and to make predictions on processes occurring in systems that have not yet studied, it is expected that this model can be extended to collisions involving molecular targets. This chapter is devoted to the simplest case of H_2.

6.1 Goal of the Present Study

Since the CTMC model was successful enough to reproduce at least qualitatively many experiments, even in the details, in the case of atomic targets, and to make predictions on processes occurring in systems that have not yet studied, it is expected that this model can be extended to collisions involving molecular targets. This chapter is devoted to the simplest case of H_2. This molecule has received a lot of attention as it may serve as a test case involving most of the complications of the molecular t + argets. However, even though this molecule is simple, the problem to treat is extremely complex, since at least 5 bodies are involved during the collision: three nuclei, including the projectile and both protons, and two electrons. In addition, the fragmentation of the target may occur, giving rise to protons whose final energies and directions depend strongly on the initial direction of the target relative to the internuclear axis and the process occurring during the collision. Finally, specific effects, related to the nature of the target, may have to take into account, such as Young-type interferences. In principle, this process cannot be studied using classical models, since these models are based on the corpuscular behavior of the particles, whereas interferences take into account their wave behavior. Nevertheless, by the way of several approximations, many results can be available and compared with experiments and other theories.

According to the desired result, two ways can be taken for H_2 (it is the same for more complex molecular targets). If the goal is to calculate total and partial cross sections, H_2 may be simply seen as an atom, with a nucleus of $Z_T = 2$, that plays the role of a reservoir of electrons. In this case, only the atomic number of the target

© The Author(s), under exclusive license to Springer Nature Switzerland AG 2021
F. Frémont, *Classical Treatment of Collisions Between Ions and Atoms or Molecules*, Springer Series on Atomic, Optical, and Plasma Physics 118,
https://doi.org/10.1007/978-3-030-89428-3_6

and the binding energies of the electrons are important. As for an atomic target, an independent electron model or a correlated electron model described previously will be used. Consequently, the problem to treat is a 4-body problem. If, on the other hand, the fragmentation of the molecule, or the effect of its initial orientation with respect to the internuclear axis, is studied, the position of the protons plays a role in the collision. It is then necessary to solve a problem with 5 bodies.

The present chapter is divided in three parts. In the first section, we focus our attention on the H_2^+ molecule, that have the great advantage to reduce the number of electrons. Then, we will consider the H_2 molecule as an atomic target in order to analyze total and partial cross sections. Finally, we will focus on the molecular aspect of H_2, by analyzing the consequences of the molecular fragmentation, in terms of final positions and energies.

6.2 Collisions Involving H_2^+ Ions

In addition to its fundamental interest, the use of H_2^+ molecular ions colliding on atomic or molecular targets is necessary to understand some specific problems as, for example, the evolution of complex chemical systems. Especially, much attention was devoted to the collision system $H_2^+ + H_2 \rightarrow H_3^+ + H$, since H_3^+ is one of the most intriguing molecular ions in hydrogen plasmas [1].

6.2.1 $H + H_2^+$ Collisions

To$^+$ our knowledge, the first classical study involving H_2^+ as a target was performed in 1998 [2]. The goal of this study was to understand how and to which degree the complex structure of the target modifies the results of the scattering with respect to the standard two nucleus-one electron case. For this collision system, the analysis was performed for $0.3 < v_P < 2$ a.u., corresponding to 2.25 keV $< E_P <$ 100 keV.

The presence of two nuclei is clearly seen when the differential capture cross section $d\sigma_{SC}/d\theta_e$ is plotted as a function of the azimuthal angle θ_e (Fig. 6.1). Especially at low velocities, the increasing trend of $d\sigma_{SC}/d\theta_e$ is visible, meaning that the orientation of the molecule prior collision has a great influence in the capture probability. Indeed, when the molecular is perpendicular to the internuclear axis, the capture probability is enhanced since capture may occur at larger impact parameters than that in the case of molecules parallel to the internuclear axis.

CTMC capture cross sections are represented in Fig. 6.2 (full red curve) as a function of E_P. Since no experimental data are available for this system, the present calculations are compared with several results derived from scaling laws [3–5] (dashed, short dashed and dotted dashed blue curves) and cross sections for the collision system $H^+ + H$ [6] (short dashed curves). It is obvious that none of these calculations reproduce the present calculations. Nevertheless, the behavior is seen to be similar,

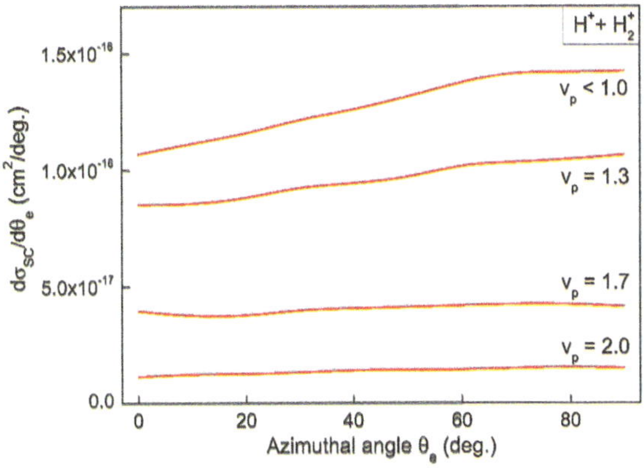

Fig. 6.1 Differential SC cross section $d\sigma_{SC}/d\theta_e$ following $H^+ + H_2^+$ collisions as a function of the azimuthal angle θ_e, for several impact velocities v_P [2]. CTMC calculations, full red curves

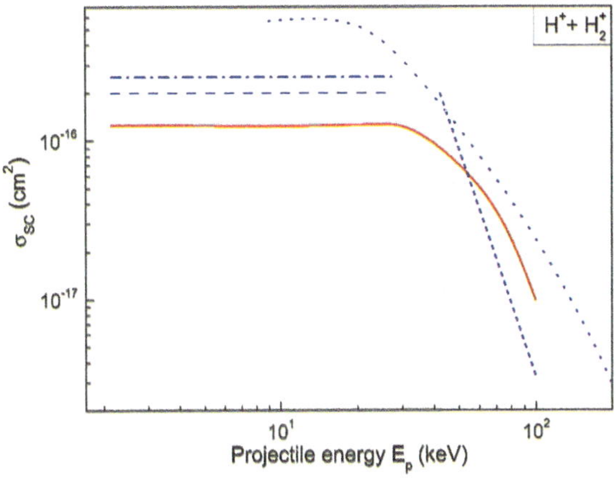

Fig. 6.2 SC cross sections following $H^+ + H_2^+$ collisions as a function of E_P. CTMC calculations, full red curve [2]; scaling laws, dashed, dotted dashed and short dashed curves [3–5]; CTMC calculations for the collision system $H^+ + H$, short dotted curve [6]

since SC cross sections were found to be constant below 30 keV and to strongly decrease above. In addition, the CTMC method used microcanonical distributions for the momentum, inducing a poor reproduction of the spatial distribution.

The relative correct agreement is an indication that a classical method is again a good tool, even for molecular targets.

6.2.2 $H + H_2^+$ Collisions

In 2005, collisions between H_2^+ ion projectiles with H targets were investigated in the 2.5 keV–1 meV energy range by means of the CTMC method [7]. The advantage of this collision system was that the target electron-projectile electron interaction could be taken into account. As in the previous case, the initial conditions for H were sampled from microcanonical distributions. With an initial H_2^+ electron velocity

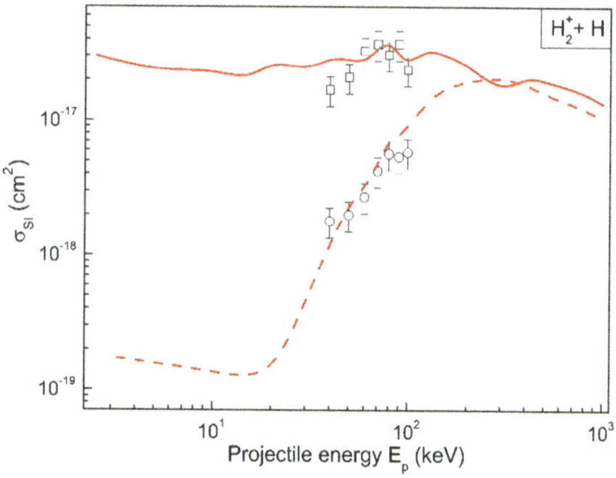

Fig. 6.3 SI ionization cross sections following $H_2^+ + H$ collisions as a function of E_P. Projectile ionization without target ionization, experiment (empty squares), CTMC calculations (full red curve); projectile ionization with target ionization, experiment (empty circles), CTMC calculations (dashed red curve)

of 1.0566 a.u., the calculated bond length and vibrational period were 0.1109 nm and 1.502 10^{-14} s, respectively, which are very close to the experimental values of 0.1070 nm and 1.45 10^{-14} s. The CTMC results were compared with previous experimental data [7].

Experimental projectile ionization without (empty squares) and with (empty circles) are shown in Fig. 6.3 as a function of E_P. The CTMC calculation for each process (full and dashed red curves) were found to agree fairly well with experiment, under the uncertainties, despite of all the approximations made in the calculations.

The agreement demonstrates that a classical representation of the H_2^+ ion is feasible without any introduction of non Coulombic model potentials. Of course, because of limitation in the experimental energy range, an extension of this range would have been desirable. Finally, the role of electron correlation would have been possible to study by neglecting it in the calculations.

6.3 Collisions Involving H_2 Molecular Targets

6.3.1 Total and Partial Cross Sections

6.3.1.1 Total Cross Sections

The first CTMC treatment of ion-H_2 collisions was performed in 1989, i.e. at least twenty years after calculations involving target atoms [8]. Two approximations were made: The projectile was assumed to follow a straight line all along the collision, and both target interacted using static model potentials, written as

$$V(r) = \frac{1}{r}\left[1 - (1 + 1.166r)e^{-2.332r}\right] \tag{6.1}$$

The electronic initial positions and momenta were sampled from a combination of two electron phase-space distributions, that resemble the quantum–mechanical ones, as seen in Figs. 6.1 and 6.2 (Figs. 6.4 and 6.5).

The cross section for a specific process P was defined as

$$\sigma_P = \frac{N_P}{N}\pi b_{max}^2 \tag{6.2}$$

where N_P and N are the number of events that lead to P and N is the total number of events, respectively. It has to be noted that this expression differs from the more accurate expression that incorporates the dependence of N_P onto b. In order to compare with experiments, cross sections were calculated as a function of E_P for different charge states of the projectile. The results for free electron loss (process

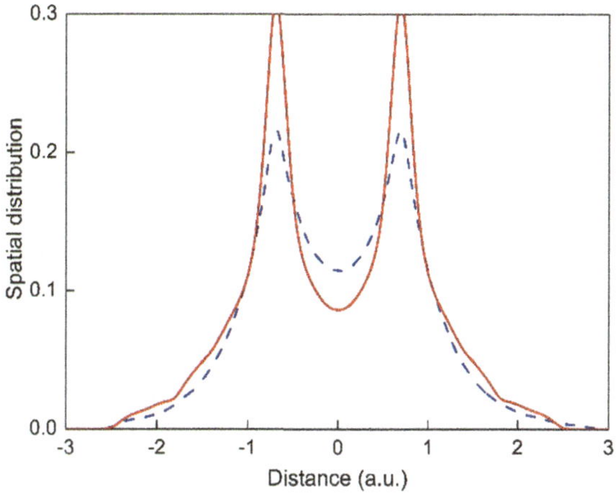

Fig. 6.4 Spatial distributions of an electron in the field of the two protons of the molecule H_2. Quantum calculations, dashed blue curve; CTMC calculations, full red curve [8]

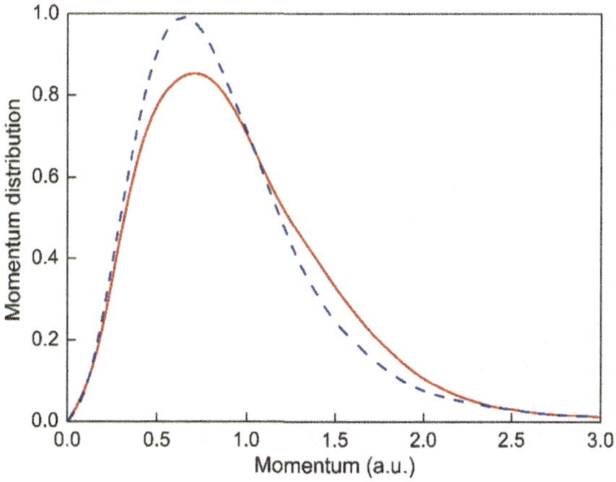

Fig. 6.5 Momentum distributions of an electron in the field of the two protons of the molecule H_2. Quantum calculations, dashed blue curve; CTMC calculations, full red curve [8]

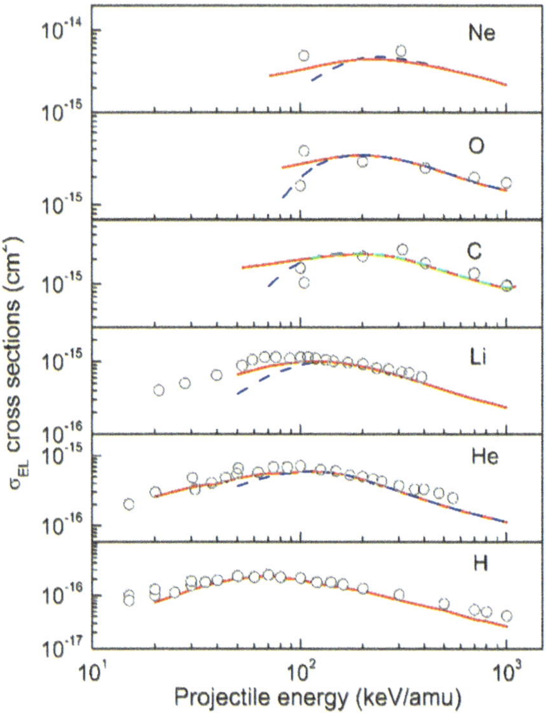

Fig. 6.6 Free electron loss (including SI, DI and TIbut not electron capture) cross sections as a function of the projectile energy in keV/amu, following A^{Z+} + H_2 ($Z = 1 - 10$) collisions. Experiment (empty circles); CTMC calculations without inclusion of autoionization, dashed blue curves; CTMC calculations with inclusion of autoionization, full red curves [8]

that includes SI, TI and DI) cross sections are displayed in Fig. 6.6. The CTMC calculations were performed without (dashed blue curves) and with (full red curves) inclusion of autoionization, and compared with various experiments (empty circles). As in the case of ions impacting on *H*, the agreement between experiment and calculations was found to be remarkable in the energy range explored. The calculations could also demonstrate that autoionization could not be neglected, especially at the lower energies.

Ten years later, the collisions systems $H^+ + H_2$ and $Be^{4+} + H_2$ were revisited, in order to compare several theoretical methods [9]. Unfortunately, for $H^+ + H_2$, no improvement was observed at velocities smaller than 1 a.u., and calculations for $Be^{4+} + H_2$ were only made at intermediate velocities, which prevents having a more precise idea of the domain of validity for CTMC calculations. Despite the lack of results, the transition probabilities were found to be similar in shape for all the models used. As an example, Fig. 6.7 shows the quantity $bp(b)$ for $H^+ + H_2$ collisions at $v_P = 1$ a.u., using the optimized dynamical pseudostates (OPD) method (dashed blue curve and short dashed blue curves, using two different model potentials) and the eikonal CTMC method (full red curve).

The maxima are located at the same value of $b = 2$ a.u., and the widths are the same whatever the model used, confirming the validity of using classical methods.

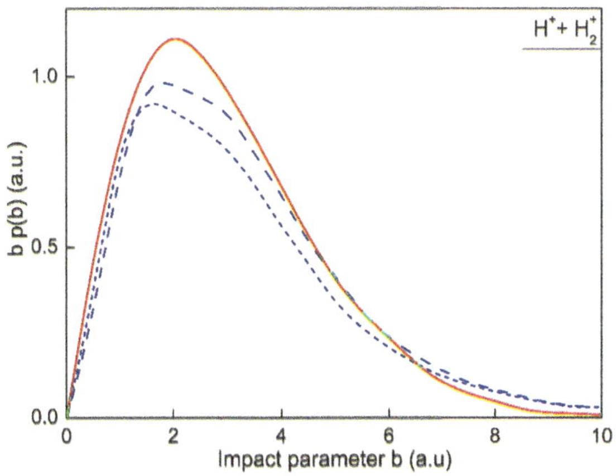

Fig. 6.7 Calculated probability distributions $bp(b)$ for $H^+ + H_2$ collisions at $v_P = 1$ a.u. OPD method, dashed blue curve and short dashed blue curves, using two different model potentials); eikonal CTMC method (full red curve)

Similar calculations (not shown here) were performed in the case of electron capture. As usual, in the energy range explored, the agreement is good between experiment and CTMC results, except in the case of the $H^+ + H_2$ system for which CTMC calculations are underestimated below $v_P = 1$ a.u.

Due to the lack of convergence of the capture cross sections for $H^+ + H_2$ collisions at projectile velocities smaller than 1 a.u., a systematic study of SC and SI was performed for this system for velocities down to 0.6 a.u., changing the potentials as seen by the active electron as well as the initial radial and momentum distributions [10], and was then extended to projectiles with higher charge states [11]. The classical treatment was based on two major approximations:

- A Franck–Condon approximation (also called sudden approximation), which prevents the molecular protons to move apart during the collision. Consequently, the cross sections were obtained by ignoring vibration and rotation of the molecule, and by keeping the distance r_{HH} between both protons fixed and equal to the equilibrium distance r_o.
- An effective Hamiltonian to describe the H_3^+ molecule. In the frame of this approximation, the interaction between the active electron and the molecule H_2^+ was described by means of an effective Coulombic potential Z_{eff}/r, with $Z_{eff} = 1.0995$. Two distributions were used. The first one was the standard microcanonical function, giving rise to a unique value of the ionization energy E. The second one was obtained by a superposition of microcanonical functions, leading to an ionization energy distribution. The energies E_i were chosen so that the mean value was equal to E. In the first case, the molecule was considered as an atom, so that the cross sections were independent onto the orientation of the molecule. In the second case, the cross sections were averaged over the orientation of the molecule.

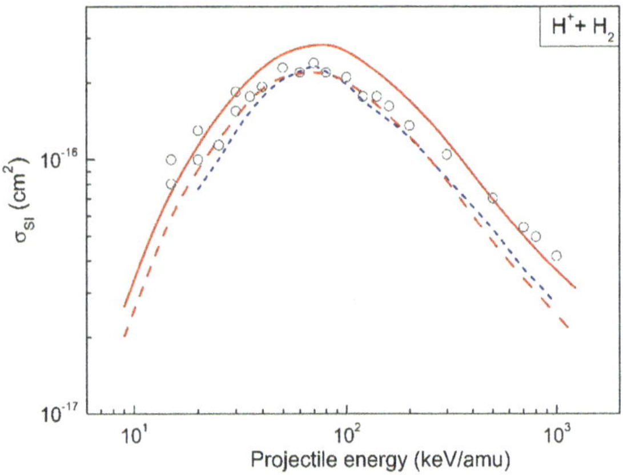

Fig. 6.8 SI cross sections following $H^+ + H_2$ collisions as a function of the projectile energy. Experiment, empty circles; CTMC calculations: Meng et al., short dashed blue curve [8]; the present work, using the atomic and average-molecular orientation approximations, dashed and full red curves, respectively [10]

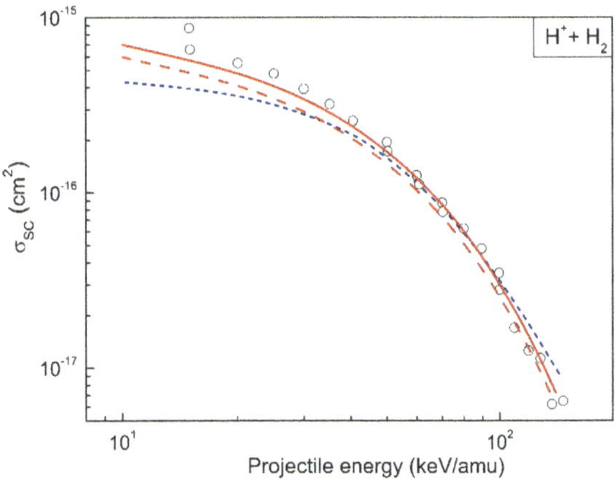

Fig. 6.9 SC cross sections following $H^+ + H_2$ collisions as a function of the projectile energy. Experiment, empty circles; CTMC calculations: Meng et al., short dashed blue curve [8]; the present work, using the atomic and average-molecular orientation approximations, dashed and full red curves, respectively [10]

The results for SI and SC processes are shown in Figs. 6.8 and 6.9 as a function of the projectile energies (dashed and full red curves using the atomic and average-molecular orientation approximations, respectively [10]), and were compared with experiment (empty circles) and the previous calculations by Meng et al. [8].

In the case of SI, the improvement is not obvious, except in the case of classical calculations using the discrete ionization energy distribution, which was in agreement with experiment at low (<60 keV) and high (>300 keV) energies, but was overestimated between these limits.

SC capture cross sections were found to be in better agreement with experiment than those found previously [8] in the whole range energies explored, demonstrating again the decisive role of the initial distributions.

Nevertheless, the agreement was poor in the case of TI process (Fig. 6.10). According to the authors, this was not surprising, since the TI process is a true two-electron process, indicating that the electron correlation play an important role.

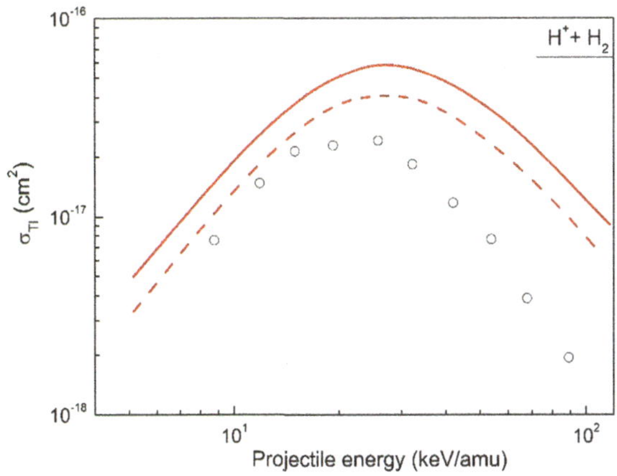

Fig. 6.10 TI cross sections following $H^+ + H_2$ collisions as a function of the projectile energy. Experiment, empty circles; CTMC calculations, present work, using the atomic and average-molecular orientation approximations, dashed and full red curves, respectively [10]

6.3.1.2 Partial Cross Sections

Surprisingly, in contrast with H targets, experimental data and CTMC calculations on state-selective electron capture on excited states of the projectile after collision on H_2 targets are very scares. Despite this lack of results, the first simplistic idea would that comes to mind would be to say that the cross sections for a given process occurring with H_2 is twice those for the same process occurring with a H target. Let's call σ_{SC}, σ_n and σ_{nl} the total cross sections and partial cross sections for populating n-states and nl-states, respectively. Thus, following the above simplification, we would expect

$$\begin{aligned} \sigma_{SC}(H_2) &= 2\sigma_{SC}(H) \\ \sigma_n(H_2) &= 2\sigma_n(H) \\ \sigma_{nl}(H_2) &= 2\sigma_{nl}(H) \end{aligned} \tag{6.3}$$

These approximations seem to be reasonable at large impact parameters b. For small impact parameters, the orientation of the molecule is expected to induce changes in the relations (6.3). The coefficient of proportionality 2 may also be changed since the ionization potential for H_2 and H are different. Indeed, when representing the experimental ratio $R = \sigma_{SC}(H_2)/\sigma_{SC}(H)$ as a function of the reduced projectile energy $\tilde{E}_p = E_p/q^{4.7}$ (empty circles in Fig. 6.11), three domains could be distinguished [8, 12].

For $\tilde{E}_p$ smaller than 10 keV/amu, R was found to be constant and smaller than 1 (~0.76). The authors explained the smallness of R as being at least due to the large potential energy of H_2. For $\tilde{E}_p$ larger than 100 keV/amu, R was also found to be constant, and of the order of 3.8, within the experimental uncertainties (not shown in Fig. 6.11). In the intermediate range, R increases when increasing $\tilde{E}_p$. The ratio calculated with the Bohr-Lindhard model (dashed blue curve) is in agreement

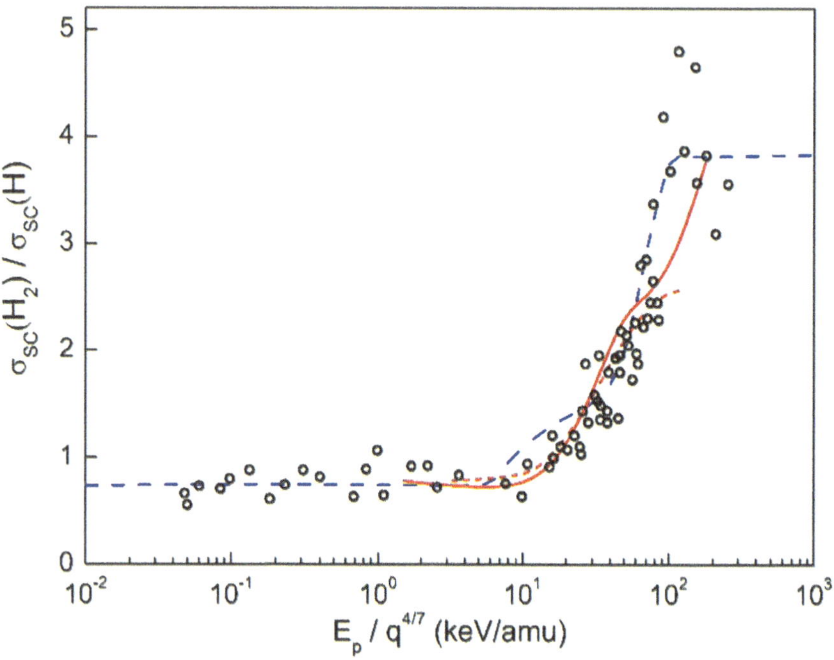

Fig. 6.11 Ratio $R = \sigma_{SC}(H_2)/\sigma_{SC}(H)$ for single capture following collisions of ions with molecular and atomic hydrogen targets, as a function of the reduced projectile energy $\tilde{E}_p = E_p/q^{4.7}$. Experiment, empty circles [12]; Calculations: Bohr-Lindhard model, dashed blue curves, CTMC model, for projectile charges $q = 2$ and 6, dashed and full red curves, respectively [8]

with the experiment, using adequate parameters for H and H_2. In Fig. 6.11 are also represented the results of the CTMC model for projectile charges q equal to 2 and 6 (dashed and full red curves, respectively). The agreement with experiment is also good. However, data are missing at low and large values of $\tilde{E}_p$. As a conclusion, the differences observed according to the energy obliges us to make specific calculations for the H_2 molecule.

The first theoretical study of (n, l) distributions following SC in $A^{q+} + H_2$ collisions was performed in 1990 [8], together with that following A^{q+} + He collisions. An original method was employed to take into account for the autoionization process by monitoring the final one-electron binding energies referred to the projectile. If both electrons were captured outside the ground-state bin, the autoionization process was assumed to automatically occur with a probability of 1.

As a benchmark the first collision system to be explored was $H^+ + H_2$ in the projectile energy range 10–120 keV. Figure 6.12 reports total and partial SC cross sections as a function of E_p. Experimental results (open circles and full squares were derived from [13]). Except for total cross sections (top of Fig. 6.12) and partial cross sections σ_{4s} for which clear discrepancies were seen, the agreement for σ_{nl} is very

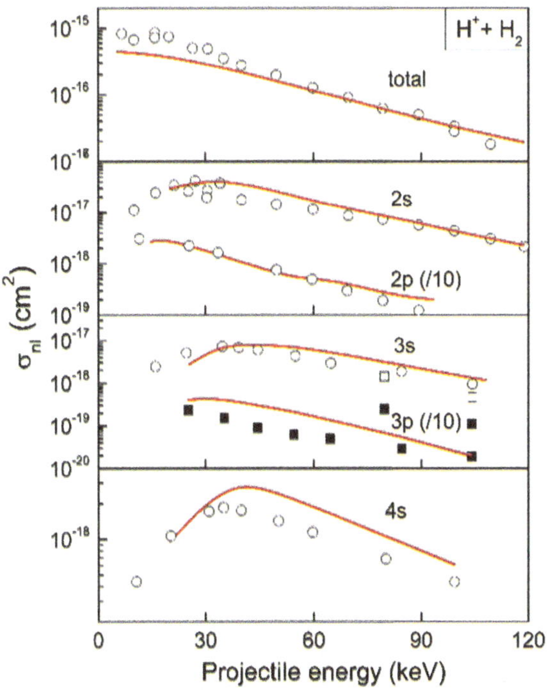

Fig. 6.12 (n, l) distributions following $H^+ + H_2$ collisions as a function of the projectile energy. Experiment, empty circles and full squares; CTMC calculations, full red curves

good, in shape and in magnitude. For the $3p$ state, no conclusion could be made, since the CTMC results lied between two different sets of data that disagreed with each other.

The CTMC calculations were extended to higher values of q. However, it was impossible to test the validity of the classical model, since no experimental data were available. Nevertheless, a complete theoretical analysis of the n distribution could be done for the collision system $Au^{14+} + H_2$ at $E_p = 100$ keV/amu, and the theoretical results were compared with the experimental distribution. The experimental (full circles) and CTMC (red curves) results are depicted in Fig. 6.13 as a function of n.

The fact that only four experimental values were reported is not surprising. It is only due to the restrictive conditions inherent to the experimental method. On the theoretical side, if autoionization (AI) was not taken into account, the theoretical distribution, given mainly by the SC process, would reflect in shape the experiment even though, quantitatively, the CTMC results were overestimated by a factor of at least 2. When the AI process was included, as seen in the Fig. 6.13, the discrepancies between experiment and calculations was much larger (at least a factor of 4). So far, no convincing reason could be invoked to explain such differences.

In 1996, a systematic study classical study was engaged to determine cross sections for state-selective excitation and electron capture following $Be^{q+} + H_2$ collisions ($q = 2$–4) at E_p ranging from 1 keV/amu up to 1 meV/amu [14]. The choice of beryllium was motivated by its importance in fusion reactors research and its presence

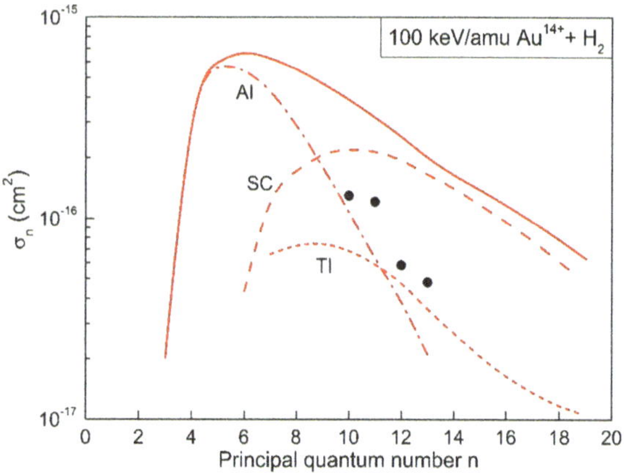

Fig. 6.13 *n* distribution following SC in 100 keV/amu $Au^{14+} + H_2$ collisions [8, 13]. Experiment, full circles; CTMC calculations: TI, short dashed curve; SC, dashed curve; AI (autoionization), dotted dashed curve; total, full red curve

in species used in fusion devices. Because of the extreme difficulty to implement quantum calculations to derive *nl* distributions, the CTMC method was the only one available to give rapid results with a fairly good precision. The problem was that no experimental result was available to compare with the present theory for such collision systems, so that the huge database constituted by the authors seems to be difficult to use in the future, despite all the efforts made. Indeed, when comparing the CTMC calculations with AO-MO quantum calculations, which are available only at low E_p, discrepancies were observed, whatever the projectile charge q (Fig. 6.14), especially for Be^{4+}projectiles.

With the development of new innovative experimental techniques, such as the cold-target recoil-ion momentum spectroscopy (COLTRIMS) technique, a new perspective for the exploration of collision processes was opened, giving rise to the determination, at a high detailed level, of state-selective SC cross sections in ion-atom and ion-molecule collisions. Recently, the COLTRIM technique was used to study the $He^+ + H_2$ collision system at intermediate impact energies, and to explore the SC at the impact parameter level [15]. Even though the SC process seems to be simple for this system, it is in fact very complicated since SC may be accompanied by ground state dissociation (GSD), target ionization (TI) or target excitation (TE). The interactions among particles were represented by Coulombic potentials, whereas the initial state of the molecular target was described by two independent H atoms hold together by a Morse potential.

First, to check the validity of the model, total SC cross sections were calculated as a function of E_p (full red curve of Fig. 6.15) and compared with experiments (empty circles, squares and triangles), in the energy range 30–600 keV. The classical calculations were found to be in excellent agreement with experiment in the energy range explored, giving thus confidence to go further to more detailed calculations.

To provide a more exhaustive analysis, the experimental and CTMC n-state selective cross sections were compared (Fig. 6.16). Good overall agreement was obtained for n varying from 1 to 3. However, CTMC calculations were underestimated for $n = 2$ at E_p smaller than 100 keV, to the advantage of calculations for $n = 2$. According

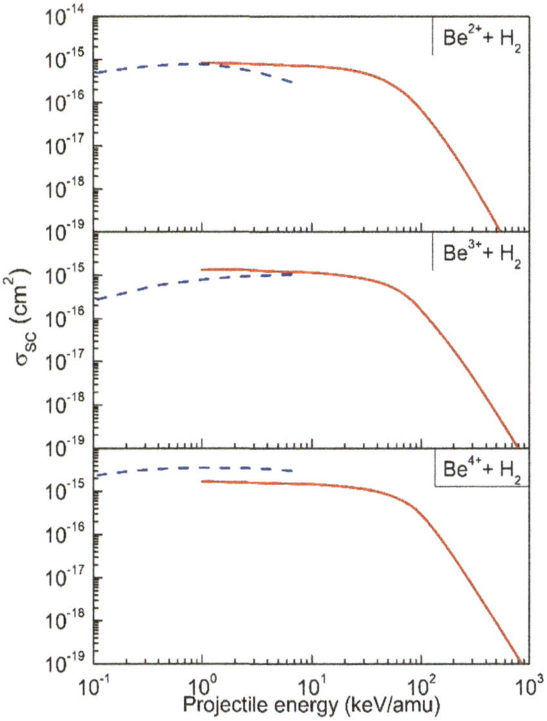

Fig. 6.14 SC capture cross sections following Be^{q+} + H_2 collisions (q = 2 – 4) at E_p ranging from 1 keV/amu up to 1 meV/amu [14]. Quantum calculations, dashed blue curves; CTMC calculations, full red curves

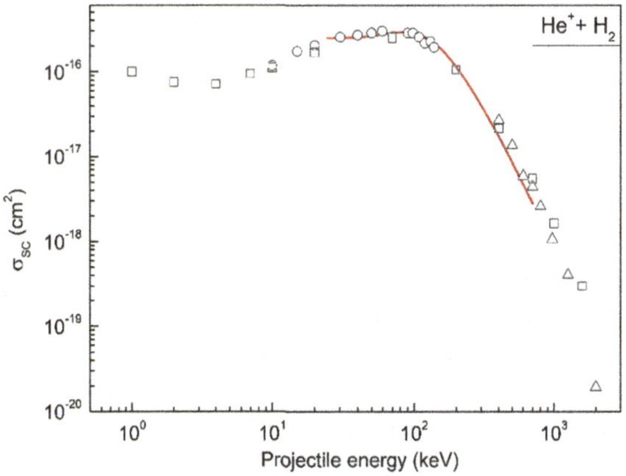

Fig. 6.15 Total SC cross sections following He^+ + H_2 collisions as a function of E_p. Experiment, empty objects; CTMC calculations, full red curve [15]

to the authors, this was probably related to the one-electron model. One can more easily think that this is due to the fact that the classical separation between n = 1 and 2 is approximate.

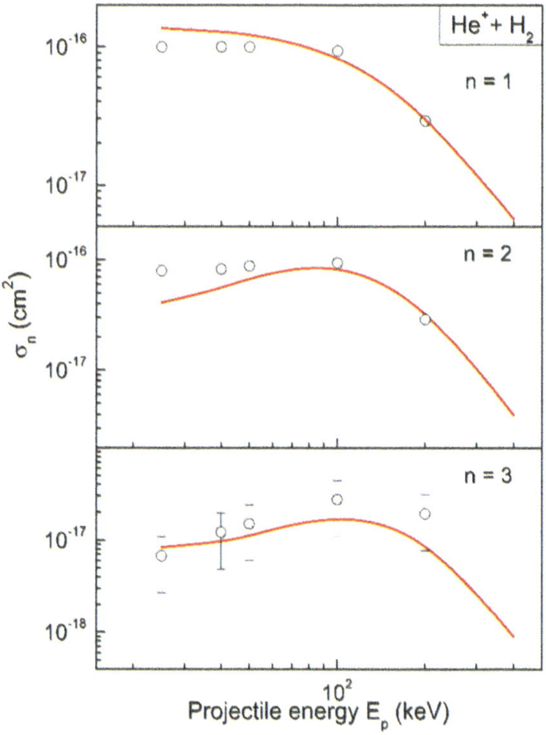

Fig. 6.16 n-state selective SC cross sections following $He^+ + H_2$ collisions as a function of E_p, for $n = 1, 2$ and 3 [15]. Experiment, empty circles; CTMC calculations, full red curves

Confidence in CTMC calculations allowed the authors to go further in detailed analysis. Let us call (xOy) the plan perpendicular to the incident projectile trajectory axis. The determination of p_y or p_x distributions give information on the impact parameter distributions. A large (respectively small) value of p_y corresponds to small (respectively large) values of b. As an example, Fig. 6.17 shows p_y distributions for $E_p = 6.5$, 10 a1.5nd 12.5 keV/amu. The maxima of the CTMC calculations (full red curves) were normalized to the experimental ones (empty circles). Whatever E_p, the experimental distributions were found to be very well reproduced by CTMC calculations for $|p_y| < 1.5$ a.u., i.e. for relatively large b. At larger values of $|p_y|$, the CTMC results were underestimated, especially at 10 keV/amu. No explanation was given by the authors was given by the authors. Nevertheless, one may invoke the fact that, at large b, the molecule resembles an atom, while the molecular character is present at small b, so that a full 5-body model would be probably useful to treat the present collision system.

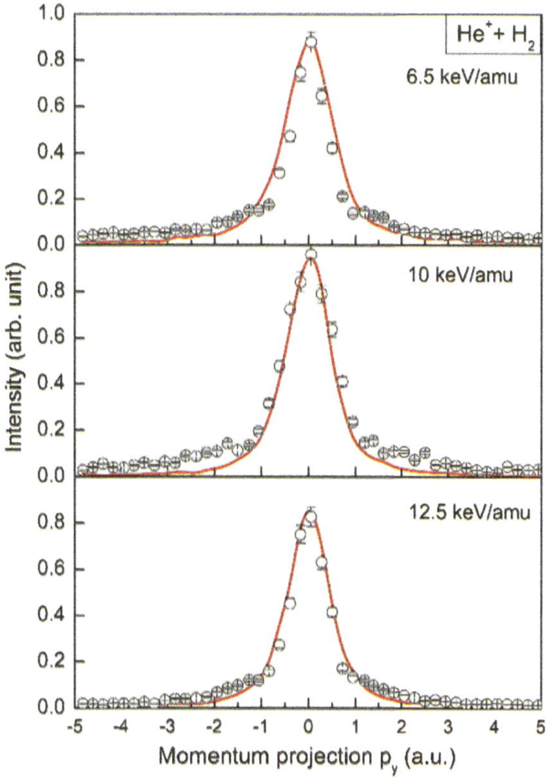

Fig. 6.17 Total experimental (empty circles) and CTMC (full red curves) momentum projection distributions along the (Oy) axis following 6.5, 10 and 12.5 keV/amu $He^+ + H_2$ collisions [15]

6.3.2 The Role of Electron Correlation

6.3.2.1 Simulation of Electron Correlation Using Independent Electron Models

It is well known that taking into consideration the electron correlation during the collision is one of the hardest tasks. To circumvent the difficulty, the electron correlation was simulated by bringing into play two variants of the CTMC one-electron model, denoted nCTMC and inCTMC [16]. The nCTMC variant treated the two electrons of H_2 as being sequentially bound, reflecting the correlation in the initial state. The second variant treated the two H_2 electrons as being bound by equal amount, reflecting an uncorrelated state. A combination of both models (75% nCTMC and 25% inCTMC) was adopted to calculate state-selective cross sections for SC in collisions of 21 keV and 42 keV $O^{6+} + H_2$. The results are presented in Fig. 6.18 (full red curves) for $n = 4$ and for l varying from 0 to 3, and compared with available experiments (empty circles and squares). It was found that the CTMC model seriously underestimated the 4s cross section by a factor of 10 compared to the experiments. This was probably due to the fact that a small value of n was involved. According

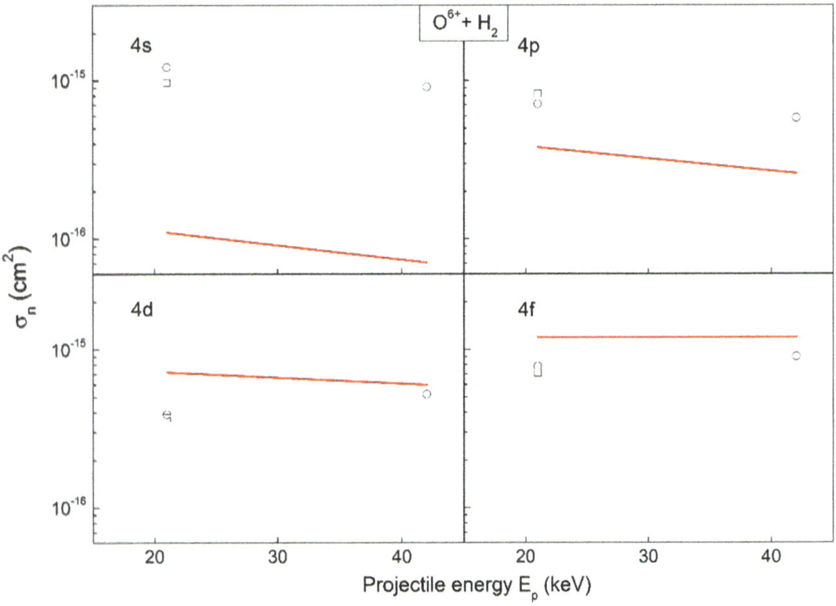

Fig. 6.18 State-selective cross sections for SC in $O^{6+} + H_2$ collisions at energies of 21 keV and 42 keV in 4 l (l = 0–3) states [16]. Experiment, empty circles and squares; CTMC calculations, full red curves

to the author, the agreement would have been better for larger n. Despite this strong discrepancy, the calculation was found to be in qualitative accord with experiment for larger values of l.

6.3.2.2 An Example of Full Calculations Using Phenomenological Potentials

The full description of H_2, including electron correlation, is in principle impossible when using only Coulombic potenatials. The inclusion of phenomenological potentials V_H, that partly describe the Heisenberg uncertainty principle (for more details, see Chap. 2) was found to be necessary to introduce the electron correlation in the initial and final states. Two works were devoted to the application of the quasi classical model in the analysis of antiproton $\overline{p}$ capture by H_2 to form $p\overline{p}$ states [17, 18] at very low projectile energies. Such collisions were first studies by Cohen [17] who focused on an extension of the Kirshbaum-Wilets (KW) model to the H_2 molecule, adding multiple additional pseudo potential V_H terms to improve the internuclear distance R_{HH} dependence of H_2 and $H_2{}^+$ potentials.

Ten years later, the same collision system was revisited [18]. The goal was to return to the fundamentals of the KW method, and to show that the previous additional terms by Cohen were unnecessary to achieve a correctly bound ground state. To do that, the α and ξ parameters of the pseudopotential V_H were modified. Instead of taking 5 and 0.96, respectively, the best values were found to be 1 and 0.894. For this pair of values, the equilibrium value R_{eq} of R_{HH} was 1.574 a.u. instead of 1.42 a.u., and the first ionization potential was identical to the experimental one. Moreover, the H_2^+ ionization energy was found to be 0.79 a.u., which is fairly close to the experimental value of 0.6 a.u.

The model was first applied to proton stopping power (Fig. 6.19) in order to figure out the pertinence of the model. When comparing with experiment (empty circles), it is clearly seen that the best value for α is 1 (full red curve), in the range from 10 up to 10^5 keV/amu, whereas for $\alpha = 2$ and 4 (dashed and dotted-dashed red curves, respectively), the model noticeably deviates from the data.

In a second step, the model proposed by Beck et al. was compared with that by Cohen, by calculating and representing capture cross sections for proton capture by antiproton, at E_p varying from 10^{-5}up to 10^{-1} keV (Fig. 6.20). Both models give

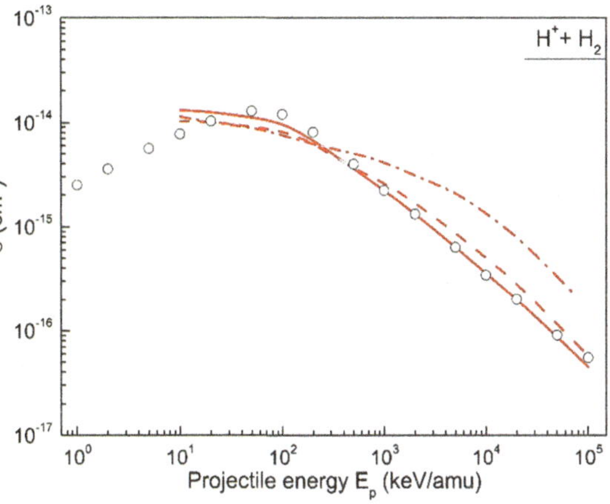

Fig. 6.19 Stopping power cross sections following $H^+ + H_2$ collisions as a function of the projectile energy [17]. Experiment, empty circles; CTMC calculations using values of α of 1, 2 and 4, full, dashed and dotted-dashed red curves, respectively

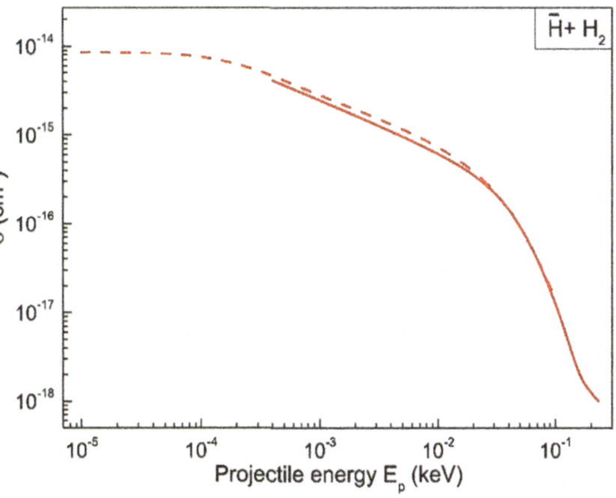

Fig. 6.20 Five-body CTMC calculations for the formation of protonium in collisions of antiprotons with molecular hydrogen [18]. Adapted model from Cohen, dashed red curve; KW model, full red curve

similar results, even though the model proposed by Beck (full red curve) was not used for E_p smaller than 4 10^{-4} keV.

As a conclusion, this model gives a real opportunity to extend the one-electron models to a model that takes into account electron correlation, and thus to quantify the importance of this process.

6.3.3 Differential Cross Sections

A complete study of SDCS and DDCS was performed for H^+ ions colliding on hydrogen atoms [19] and molecular hydrogen [20]. In the case of H_2 targets, no CTMC calculations were developed, while they were done for H targets. Therefore, to compare calculations with experiment, the latest results were divided by a factor of 2, in order to take into account the number of H_2 electrons.

In both cases, angular and energy distributions were systematically studied and compared, and results for H targets were compared with several theories, such as the first Born approximation (FBA) model, CDW-EIS approximation, and the one-electron CTMC model. Figures 6.21 and 6.22 show SDCS $d\sigma_I/d\Omega$ and $d\sigma_I/dE_e$,

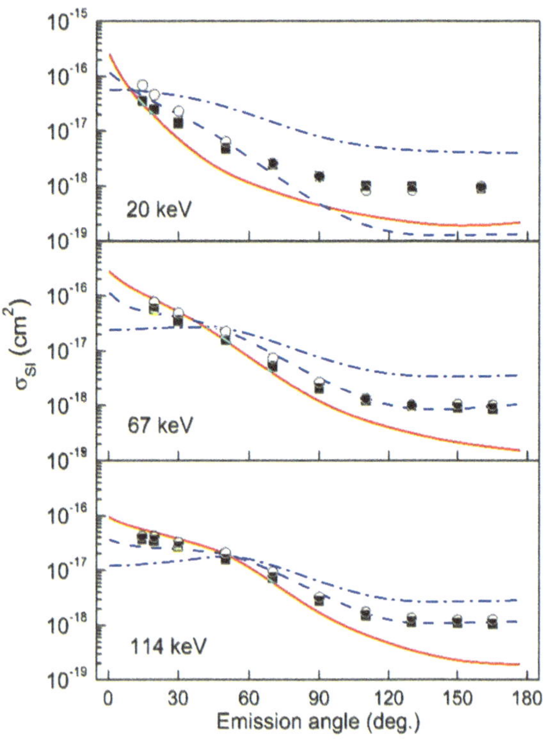

Fig. 6.21 Angular distributions following SI in $H^+ + H$ and $H^+ + H_2$ collisions, at projectile energies of 20, 67 and 114 keV [20]. The cross sections for H_2 target are divided by 2. Experiment, H target, empty circles, H_2 target, full squares; CTMC calculations, full red curves, CDW-EIS approximation, dashed blue curve, FBA, dotted-dashed blue curve

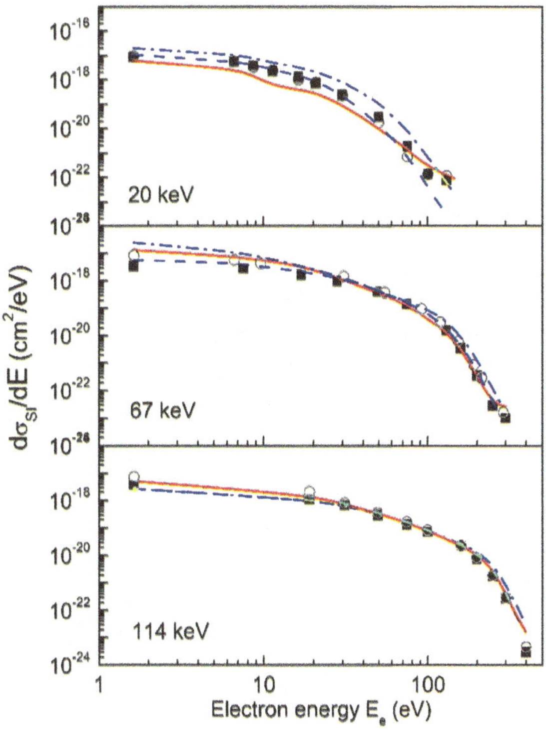

Fig. 6.22 Energy distributions following SI in $H^+ + H$ and $H^+ + H_2$ collisions, at projectile energies of 20, 67 and 114 keV [20]. The cross sections for H_2 target are divided by 2. Experiment, H target, empty circles, H_2 target, full squares; CTMC calculations, full red curves, CDW-EIS approximation, dashed blue curve, FBA, dotted-dashed blue curve

respectively, for collisions energies of 20, 67 and 114 keV. It is noted that cross sections for H_2 ionization are close to those for H, after division by 2, especially in the electron energy distribution. In the angular distributions, the overlap is also remarkable, except at angles smaller than 60° at the lowest energy E_P.

The FBA matched the experiment only in the energy distribution (Fig. 6.22), and at the highest velocity. In the other cases, the FBA model was not able to reproduce the data. The CTMC and CDW-EIS models reproduced the experimental angular distributions. However, the CTMC calculations strongly underestimated the data at angles larger than 60° by a factor that can reach 10. The experimental energy distributions was found to be well reproduced by the three models, except at 20 keV where deviations were observed when using quantum calculations.

From these results, it is clear that more theoretical analysis using the CTMC method would be of great interest, especially to check the validity of the initial conditions and, again, the role of electron correlation.

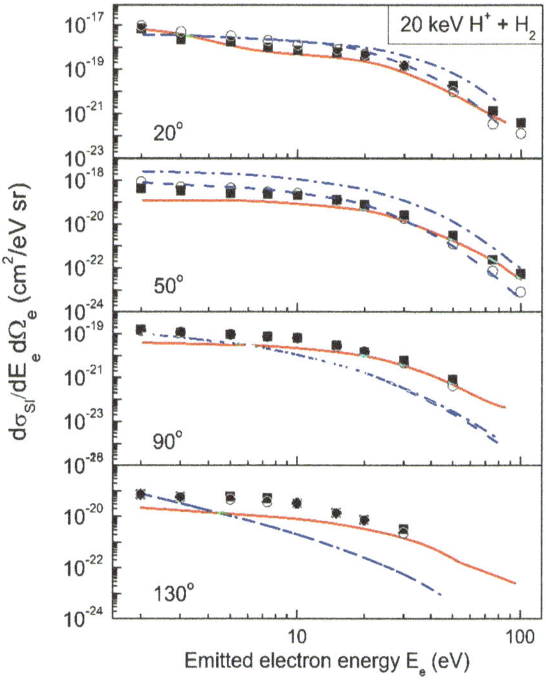

Fig. 6.23 DDCS following SI in 20 keV $H^+ + H$ and $H^+ + H_2$ collisions, at electron emission angles of 20, 60, 90 and 130° [20]. The cross sections for H_2 target are divided by 2. Experiment, H target, empty circles, H_2 target, full squares; CTMC calculations, full red curves, CDW-EIS approximation, dashed blue curve, FBA, dotted-dashed blue curve

The analysis of DDCS was also a good test for the comparison of experimental and theoretical results, especially at low energies where the CTMC model is expected to fail.

The experimental DDCS for H (empty circles) and for H_2 (full squares), divided by the same factor of 2, are presented in Fig. 6.23 at an energy of 20 keV for detection angles of 20, 50, 90 and 150°. The FBA model yielded too small cross sections (dotted-dashed curves). On the contrary, the CDW-EIS (dashed curves) and CTMC (full curves) calculations reproduced quite well the experiment, except at low electron energies. When increasing E_P, the three theories were found to converge and to reproduce the experiments, except at small angles. It is clear that the introduction of the H_2 target in the CTMC calculations is desirable and necessary.

Finally, TDCS calculations were performed in order to compare with available experimental data obtained using multi coincidence detections. A 3B-CTMC model, similar to that performed for $e^- + He$ collisions, was used to describe SI following $e^+ + H_2$ collisions at $E_P = 50$ eV and 100 eV [21]. The goal was to show that all the particles, including the two protons, actively participated in the ionization process.

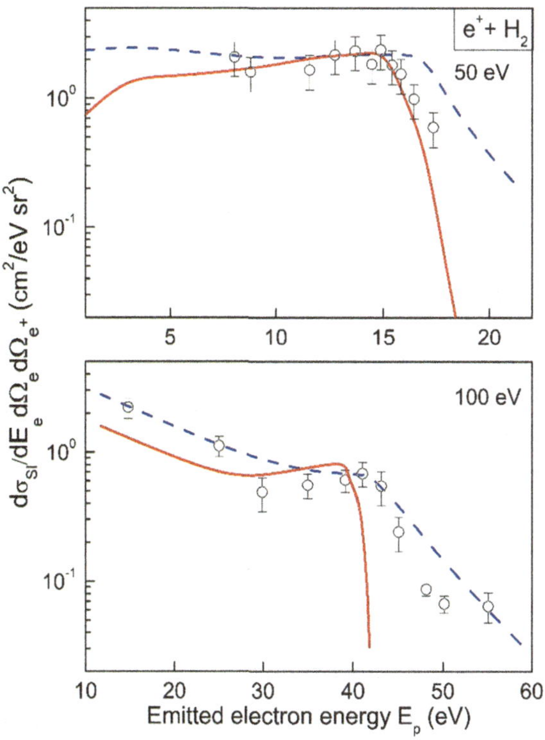

Fig. 6.24 TDCS following SI in $e^+ + H_2$ collisions at $E_P = 50$ eV and 100 eV, as a function of the emitted electron energy [21]. Experiment, empty circles; CDW calculations, dashed blue curve; CTMC calculations, full red curves

The experimental TDCS for the projectile e^+ and the electron e^- emitted at forward angles, where the ECC (electron capture to the continuum process) can be revealed, was represented (Fig. 6.24) as a function of the electron energy (empty circles) and compared with CTMC and CDW calculations (full red and dashed blue curves, respectively). The quantum mechanical approximation gave a better description of the data at 100 eV (bottom of Fig. 6.24). However, a shift in the position of the ECC peak was observed at 50 eV (top of Fig. 6.24). The CTMC results gave a better agreement with the experiment at 50 eV, but presented also a shift of the maximum at 100 eV. In addition, at both projectile energies, the CTMC cross section was found to fall down for E_P larger than 40 eV. Nevertheless, despite these discrepancies, the CTMC method was able to describe in details the SI process, and then was used to reveal the role of the collision partners during this process.

6.3.4 *Molecular Target Fragmentation*

6.3.4.1 Interest of the Study

When analyzing in energy or in angle the emitted electron or the projectile, many problems occur. At low energy, typically 1–10 eV, the electrons are very difficult to be detected, due to residual magnetic fields for example, or inhomogeneity of the applied electric fields. Similarly, the projectile gains energy, but the difference of energy before and after the collision is so small that conclusions are not obvious. In addition, the scattering angles are also generally small, of the order of 1 mrd or less, so that the angular distribution is very narrow, preventing a full analysis of the different channels involved in the collision.

In contrast, the detection of fragments is a unique opportunity to analyze in details a collision involving a molecular targets, since the fragments are, in principle, emitted in all the directions. Moreover, when fragmentation occurs, it leads to fragments H or H^+ whose energies are larger than 1 eV, giving the possibility to detect them easily.

6.3.4.2 Fragmentation of H_2 Following DC Using Xe^{23+} and O^{5+} Ions

The first experimental analysis of molecular fragments was performed at the end of the 1990's. The work was focused on the projectile charge dependence of the fragment energy distributions [22]. The projectiles Xe^{23+} and O^{5+} were used at projectile energies of 345 keV and 75 keV, respectively, corresponding to velocities of 0.33 and 0.43 a.u., respectively. The fragment energies were analyzed at detection angles θ_d ranging from 20° (forward angles) to 160° (backward angles), with respect to the incident projectile direction. The use of two different projectile charges q was motivated by the fact that q is expected to influence the energy and the direction of the emitted fragments. This effect is similar to the well-known post collision interaction (PCI) effect called Barker and Berry effect, which influences the energy spectra of Auger electrons emitted from the projectile in the field of the ionized target.

To proceed with the analysis of the experimental results, the energy distributions of H^+ ions were calculated using a two-step two-body (2S–2B) and a one-step three body (1S–3B) model. The first model supposed that the fragmentation occurs after the double capture, while the second model treated the two protons and the residual projectile ion simultaneously. Here, we concentrate on the latter model, which is a variant of the CTMC model. Indeed, the present model assumed a sudden DC process with a probability of 1, occurring for impact parameters varying from 0 to $b_{\max}$. Thus, the problem to solve was a pure 3B problem involving Coulombic interactions between the projectile and the protons.

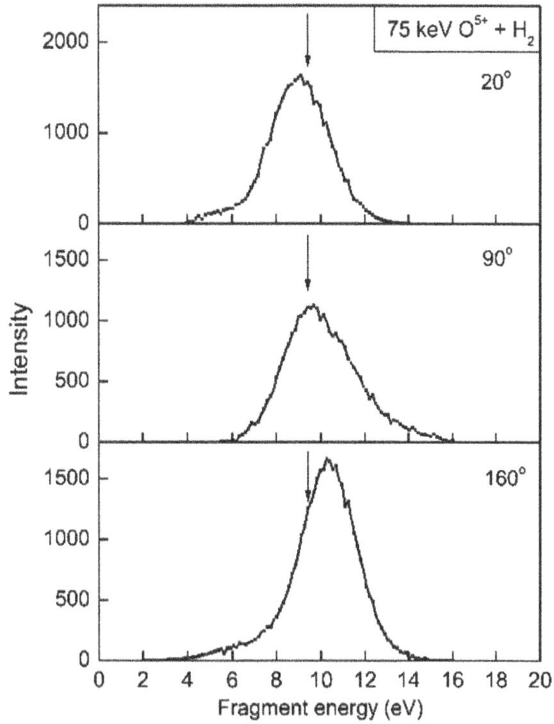

Fig. 6.25 Experimental energy distributions of the H^+ ions following DC in O^{5+} + H_2 collisions, at $\theta_d = 20°$, 90° and 160°. The arrow indicates the maximum of the distribution in the case of a pure Franck–Condon transition

Figure 6.25 presents typical experimental distributions of H^+ ions following DC in $O^{5+} + H_2$ collisions, at $\theta_d = 20°$, 90° and 160°. In the case of a pure Franck–Condon (or vertical) transition, the energy of the fragments would be centered at ~ 9.5 eV (indicated by the arrows in the Fig. 6.25). Due to the influence of the projectile, the structures were found to be shifted to lower (resp. larger) energies at forward (resp. backward) angles.

In the case of $Xe^{23+} + H_2$ collisions (Fig. 6.26), the effect of the projectile charge is more pronounced than that observed above. At 20°, the maximum structure is located at 6.8 eV, whereas it is located at 12.9 eV at 160°. In addition, at 90°, a double structure was observed, and explained in terms of kinematical addition of velocity vectors.

The experimental mean energies of the fragments were determined by fitting the experimental spectra with Gaussian curves. They were compared with the mean energies using the 3B–1S classical model (Figs. 6.27 and 6.28 for $O^{5+} + H_2$ and $Xe^{23+} + H_2$ collisions, respectively). The agreement between experiment and calculations was remarkable, indicating the strong influence of the projectile, since it is taken into account in the calculations.

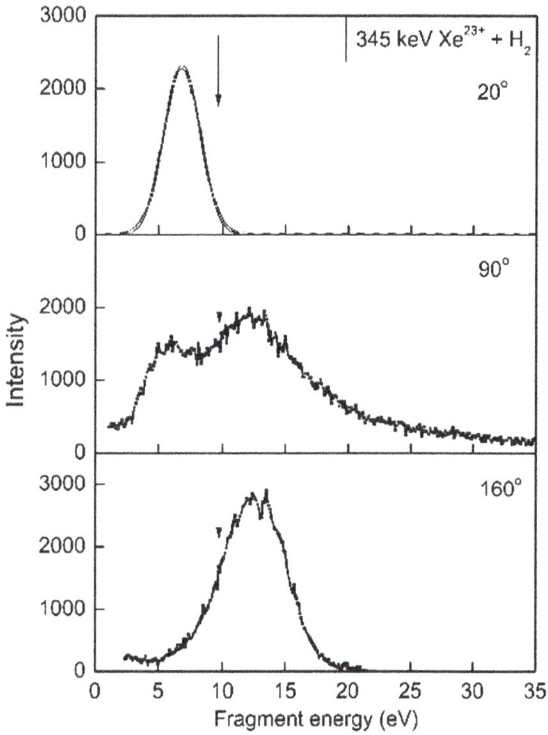

Fig. 6.26 Experimental energy distributions of the H^+ ions following DC in $Xe^{23+} + H_2$ collisions, at $\theta_d = 20°$, 90° and 160°. The arrow indicates the maximum of the distribution in the case of a pure Franck–Condon transition

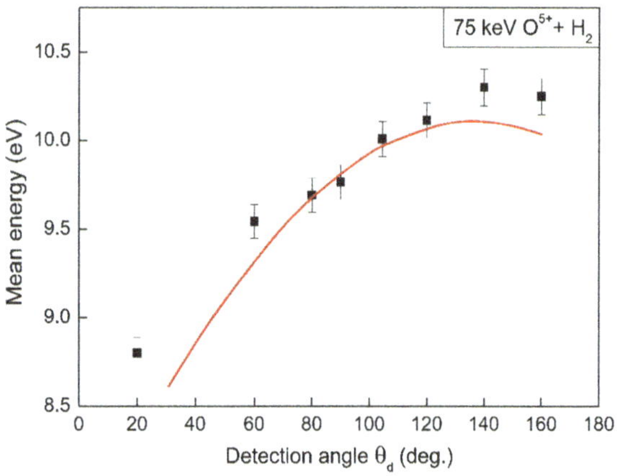

Fig. 6.27 Mean energy of proton fragments emitted after DC following 75 keV $O^{5+} + H_2$ collisions, as a function of the detection angle θ_d [22]. Experiment, full squares; 1S–3B calculations, full red line

A full 5B-CTMC model [23], including the two protons, the two active electrons and the projectile, was developed and applied to help understand the collisional mechanisms involved in ion impacting on H_2, by investigating several collision systems, such as $Xe^{23+} + H_2$, for which experimental results were available. The calculated proton energy distributions (full red curves in Fig. 6.29) were compared

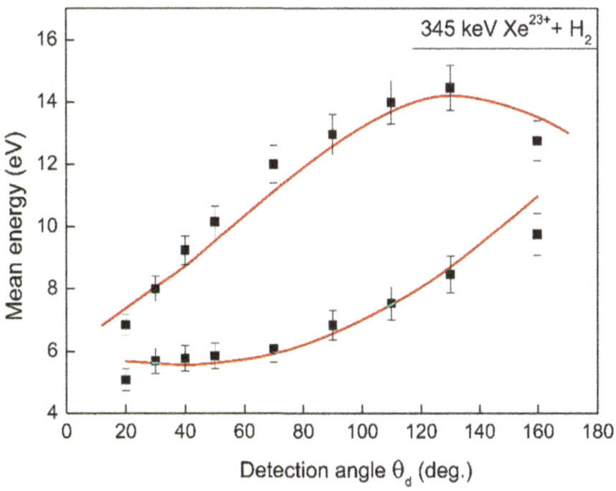

Fig. 6.28 Mean energy of proton fragments emitted after DC following 345 keV Xe^{23+} + H_2 collisions, as a function of the detection angle θ_d [22]. Experiment, full squares; 1S–3B calculations, full red line

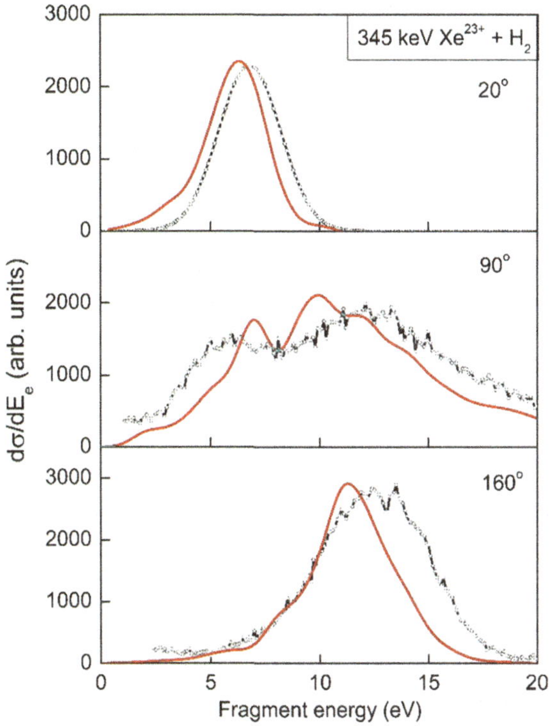

Fig. 6.29 Experimental (empty circles) and 5B-CTMC [23] (full red curves) calculated energy distributions following DC in 345 keV Xe^{23+} + H_2 collisions, for detection angles of 20°, 90° and 160°

with the experimental one presented above (Fig. 6.28), at the same detection angles. The agreement was only qualitative whatever the detection angle. Differences in the shape as well in the distribution maxima and widths were observed.

The calculations were extended to lower projectile energies, down to 2.6 eV/amu. Typical proton momentum spectra are represented in Fig. 6.30. In this configuration, the quantities p_x and p_z were defined as the components of proton momenta in the perpendicular and scattering planes, respectively. As E_P decreases, a strong proportion of H^+ ions was seen to be scattered at small angles and high momenta, indicating that capture occurred in the incoming part of the collision, before the distance of closest approach, and at very small impact parameters.

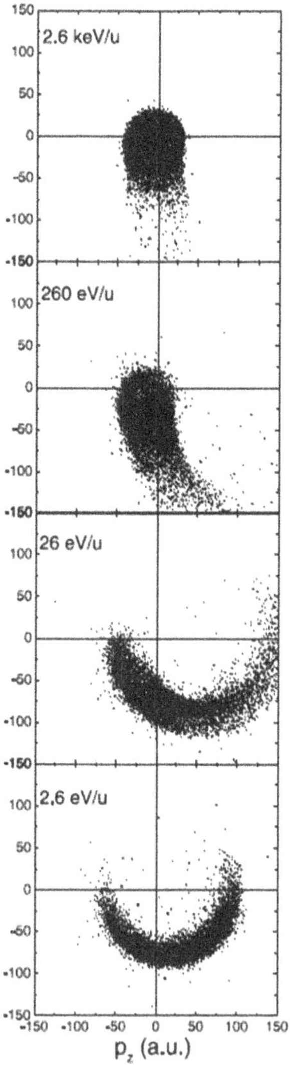

Fig. 6.30 Proton momentum distributions in the laboratory reference frame for DC in $Xe^{23+} + H_2$ collisions. The x component of the proton momenta is given in terms of the projectile scattering plane. The positive z-direction is along the projectile's incident velocity

6.3.4.3 Fragmentation of H_2 Following DC Using O^{5+} Ions at Very Low Projectile Velocities

Because calculations predicted an important fraction of H^+ emitted fragments in the forward direction at low E_P, it was necessary to compare the predicting calculations with experiment. Because it was too difficult to slow down Xe^{23+} ions, the choice fell on O^{5+} ions. The collision system $O^{5+} + H_2$ was analyzed at projectile energies ranging from 100 eV up to 105 keV, corresponding to v_P between 0.016 and 0.51 a.u. [24].

Since the most interesting DC process occurs at small detection angles θ_d, attention was focused first on the energy distribution at forward angles. Figure 6.31 shows typical spectra for $\theta_d = 30°$ at E_P of 105, 2.5 and 0.5 keV.

At the highest energy, (top of Fig. 6.31), a sharp peak, centered at ~ 9.5 eV, was observed. It was originating from the Coulomb explosion of H_2^{2+} following DC process at large b (soft collisions). As E_p decreases, the energy distribution due to soft collisions was no longer centered at 9.5 eV, but shifted to lower and lower

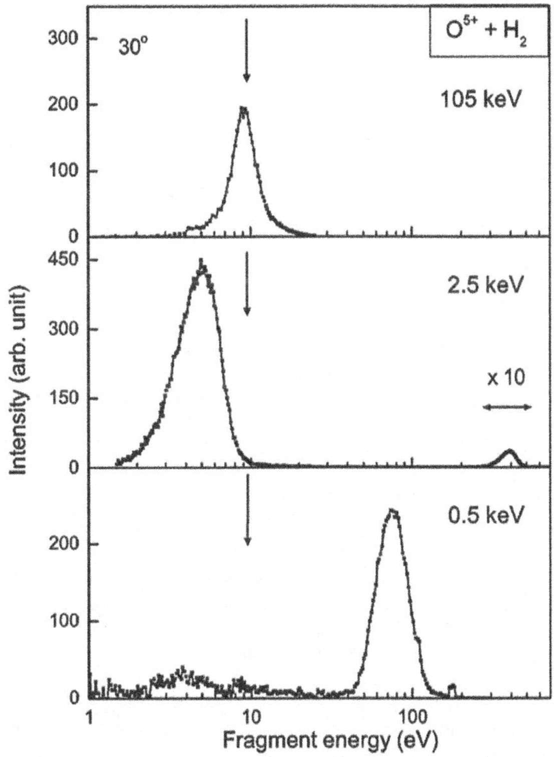

Fig. 6.31 Energy distribution of protons emitted at $\theta_d = 30°$, after DC in 105, 2.5 and 0.5 keV O^{5+} + H_2 collisions [24]. The arrow indicates the position of the maximum when a Franck–Condon transition occurs

energies. At the lowest energy, this contribution disappeared, due the repulsion of the fragments by the projectile. Another structure appeared, centered at large H^+ energies, and its intensity was found to increase with decreasing E_p. At 0.25 keV, the influence of the projectile was so strong that only fast protons were observed. Moreover, the shape of the distributions was strongly different from the one observed at 105 keV, indicating a change in the DC mechanism. The observation was coherent, at least qualitatively, with the conclusions given by the calculations [23].

Since the latter model did not give any information on the shape of the spectra, the 1S–3B model used previously was applied to the present system. The DC process was assumed to occur at b and R_{HH} (internuclear distance between the protons) that were used as free parameters, and at distance between the projectile and the target center of mass (CM) of about 7 a.u. The calculated energy distribution of the H^+ ions is presented in Fig. 6.32 as a function of θ_d for $E_P = 10$, 2.5 and 0.25 keV.

In Fig. 6.33, the maxima of the calculated energy distributions (full red curves), are compared with those obtained experimentally (full squares), for $E_P = 10$, 2.5 and 0.25 keV. At the largest energy (top of the figure), the distance R_{HH} was 1.42 a.u, i.e. the equilibrium distance. For all values of E_P, the calculations were made using the equilibrium distance (dashed blue curves) and an adapted distance (full red curves) in order to match the experiment. From the Figure, it is clear that at the

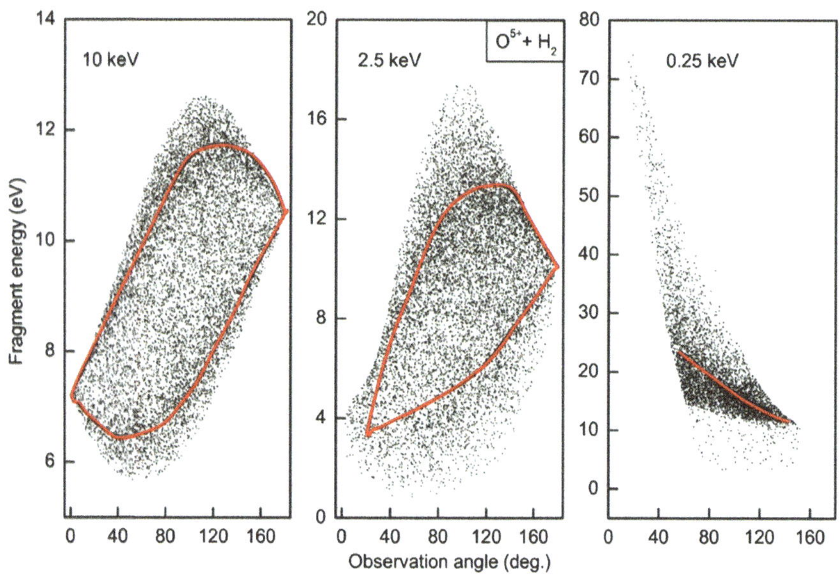

Fig. 6.32 Energy distributions, as a function of the observation angle, resulting from the model calculations, of emitted protons in DC following $O^{5+} + H_2$ collisions at projectile energies of 10, 2.5 and 0.25 keV. The maximum of the distributions (solid red curves) is also given

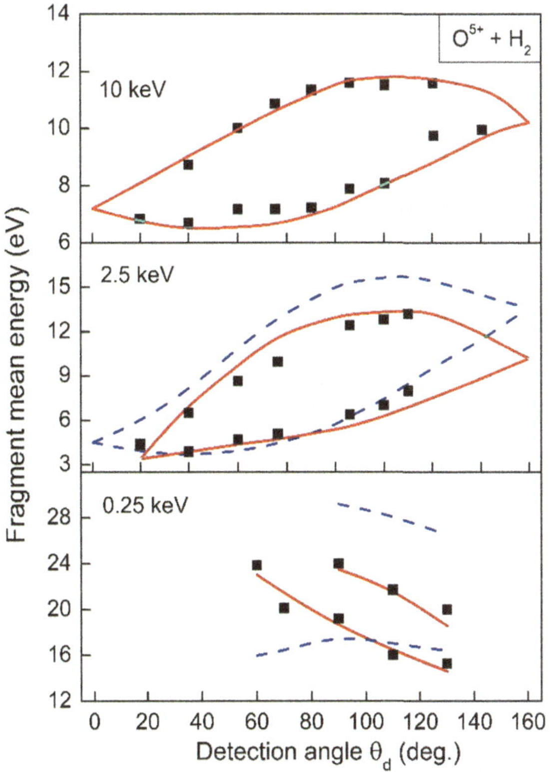

Fig. 6.33 Mean energies of H^+ fragments following DC in $O^{5+} + H_2$ collisions at projectile energies of 10, 2.5 and 0.25 keV, as a function of θ_d [24]. Experiment, full squares; 1S–3B CTMC calculations using the equilibrium distance (dashed blue curves) and an adapted distance (full red curves)

highest energy, the two curves are indistinguishable, indicating that DC occurs quite suddendly. At 2.5 keV, the projectile extends a little the molecule before the DC process, while this distance has to be much larger prior the collision at 0.25 keV in order to match the energy distributions [23].

The values for R_{HH}, taken to match the experiment, are summarized in Fig. 6.34 (full circles. The red curves is a guide for the eye) as a function of the projectile velocity. At 10 keV, the distance was 1.5 a.u., which is close to the equilibrium distance. At 2.5 keV and 0.25 keV, the distances were equal to 1.7 and 3.1 a.u., respectively. At the limit of a pure stabilized molecule $(H_2O)^{5+}$, the distance would be of the order of 4–5 a.u. It was noted that the values for R_{HH} were consistent with those found earlier (full squares) for the collision system $Ar^{5+} + H_2$ (see the reference in [24]).

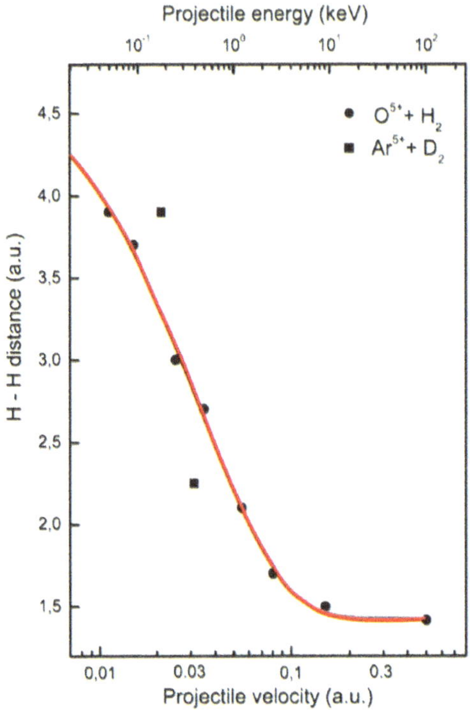

Fig. 6.34 Values of R_{HH} calculated using the 1S–3B model prior the DC process in $O^{5+} + H_2$ collisions (full circles and red curve) as a function of v_P [24]. Values of R_{HH} for $Ar^{5+} + H_2$ collisions, full squares

Since hard collisions occur at small values of b, it was assumed that, in the first approximation, the DC process was accompanied by a binary collision between the projectile and one proton, while the other fragment played a more negligible role. Thus, the low-energy collisions could be treated as a two-body problem. By analogy with binary collisions involving only two centers, applying the energy and momentum conservation rules, the experimental energy maxima (full squares) were represented (Fig. 6.35) as a function of θ_d for $E_P = 2.5$ keV and 0.25 keV, and compared with the results of binary collision model (dashed red curves). In addition, the results obtained with the 1S–3B model were also represented (full red curves). Both models gave similar results at the highest energy. It was noted that the binary model is limited to angles smaller than 90°, since the Q-value is equal to 0.

Finally, the impact parameters distributions were used at each projectile energy to determine the relative contributions of soft (dashed curves) and hard (dotted curves) collisions to the total cross sections (Fig. 6.36). At the energy of 2.5 keV (top of the

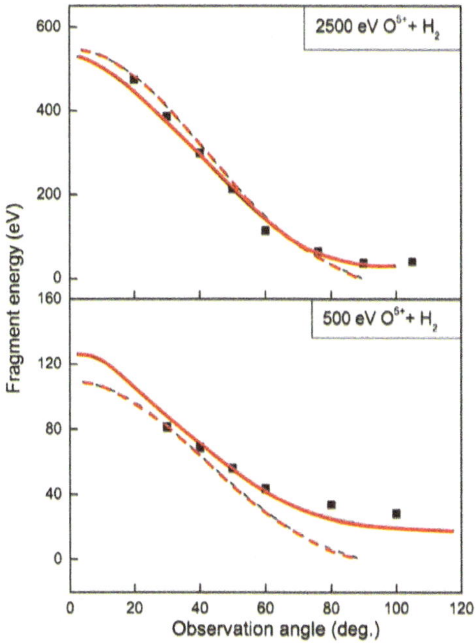

Fig. 6.35 Mean energy of a fragment emitted after a hard collision with the O^{5+} projectile after DC process, at $E_P = 2.5$ keV and 500 eV, as a function of θ_d. Experiment, full squares; Calculations: Binary model, dashed red curves, 1S–3B model, full red curves

figure), hard collisions competed with soft collision mainly at angles close to 80°. At forward and backward angles, soft collisions were dominant. Moreover, most of the protons were pushed backwards. When dividing v_P by a factor of ~3 (bottom of the figure), both contributions were of the same order of magnitude. However, soft collisions are responsible of the ejection of protons in the backward direction, while hard collisions give rise to protons which are emitted against the projectile (forward angles).

Despite the studies done on molecular fragmentation at high and low energies, there is clearly a lack of comparison between experiments and classical theory. A more complete theoretical study would be necessary, in particular by including the interaction between the electrons. We will see in the conclusions and perspectives that this introduction reveals difficulties that have not been yet resolved.

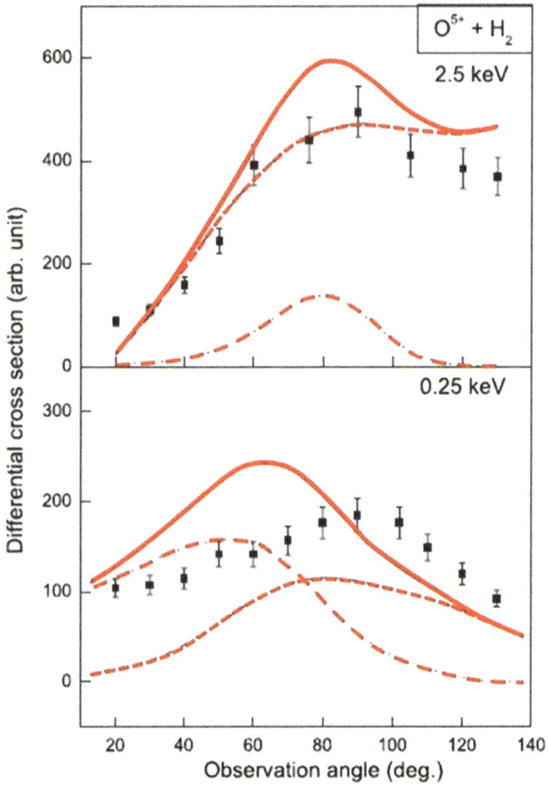

Fig. 6.36 Differential cross section in angle following DC in $O^{5+} + H_2$ collisions at $E_P = 2.5$ keV and 500 eV, as a function of θ_d. Experiment, full squares; 1S-3B calculations: contribution of soft and hard collisions, short dashed and dashed red curves, respectively; total contribution, full red curves

References

1. Savic, I., Schlemmer, S., Gerlich, D.: Formation of in collisions of with studied in a guided ion beam instrument. Chem. Phys. Chem. **21**, 1429 (2020)
2. Sattin, F., Salasnich, L.: A CTMC study of collisions between protons and molecular ions. Phys. Scrip. **58**, 464 (1998)
3. Müller, A., Salzborn, E.: Scaling of cross sections for multiple electron transfer to highly charged ions colliding with atoms and molecules. Phys. Lett. **62A**, 391 (1977)
4. Schlachter, A.S.: Charge-changing collisions. In: Proceedings of the Xth International Conference on Cyclotrons and their applications, East Lansing, Michigan USA 563 (1984)
5. Kusakabe, T., Shirai, T.: Charge transfer cross sections of slow multiply charged neon ions in collisions with noble gas atoms. Plas. Fus. Res. **9**, 3401119 (2014)
6. Illescas, C., Riera, A.: Classical study of single-electron capture and ionization processes in + (H, $_{\rm H}2$) collisions. Phys. Rev. A **60**, 4546 (1999)
7. Hennecart, D., Pascale, J.: Classical approach to collisions. Phys. Rev. A **71**, 012710 (2005)
8. Meng, L., Reinhold, C.O., Olson, R.E.: Electron removal from molecular hydrogen by fully stripped ions at intermediate energies. Phys. Rev. A **40**, 3637 (1989)

9. Elizaga, D., Errea, L.F., Gorfinkel, J.D., Illescas, C., Mendez, L., Macias, A., Riera, A., Rojas, A., Kroneisen, O.J., Kirchner, T., Lüdde, H.J., Henne, A., Dreizler, R.M.: Theoretical analysis of electron capture and electron loss in collisions. J. Phys. B **32**, 857 (1999)
10. Illescas, C., Riera, A.: Classical treatment of H+ + H2 collisions. J. Phys. B **32**, 857 (1999)
11. Illescas, C., Riera, A.: Classical study of single-electron capture and ionization processes in A^{q+} + (H, $_H$2) collisions. Phys. Rev. A **60**, 4549 (1999)
12. Knudsen, H., Haugen, H.K., Hvelplund, P.: Single-electron capture by highly charged ions collidind with atomic and molecular hydrogen. Phys. Rev. A **24**, 2287 (1981)
13. Sorensen, J., Andersen, L.H., Hvelplund, P., Knudsen, H., Liljeby, L., Nielsen, E.H.: Cross sections for electron capture collisions between medium velocity, highly charged ions and molecular hydrogen. J. Phys. B **17**, 4743 (1984)
14. Schultz, D.R., Krstic, P.S., Reinhold, C.O.: Inelastic processes in 1–1000 keV/u collisions of Be^{q+} ions with atomic and molecular hydrogen. Phys. Scr. **T62**, 69 (1996)
15. Alessi, M., Cariatore, N.D., Focke, P., Otranto, S.: State-selective electron capture in He^+ + H_2 collisions at intermediate impact energies. Phys. Rev. A **85**, 042704 (2012)
16. Machacek, J.R., Mahapatra, D.P., Schultz, D.R., Ralchenko, Y., Chutjian, A., Simcic, J., Mawhorter, R.J.: Measurement and calculation of absolute single- and double-charge-exchange cross sections for O^{6+} ions at 1.17 and 2.33 keV/u impacting He and H_2. Phys. Rev. A **90**, 052708 (2014)
17. Cohen, J.S.: Molecular effects on antiproton capture by H_2 and the states of formed. Phys. Rev. A **56**, 3583 (1997)
18. Beck, W.A., Wilets, L., Alberg, M.A.: Semiclassical description of protonium formation in antiproton collisions with molecular hydrogen. Phys. Rev. A **74**, 052707 (2006)
19. Kerby III, G.W., Gealy, M.W., Hsu, Y.-Y., Rudd, M.E.: Energy and angular distributions of electrons from ion impact on atomic and molecular hydrogen. I. 20–114 keV H^+ + H_2. Phys. Rev. A **51**, 2247 (1995)
20. Gealy, M.W., Kerby III, G.W., Hsu, Y.-Y., Rudd, M.E.: Energy and angular distributions of electrons from ion impact on atomic and molecular hydrogen. II. 20–114 keV H^+ + H. Phys. Rev. A **51**, 2256 (1995)
21. Fiol, J., Olson, R.E.: Three-body dynamics in the ionization of hydrogen by positron impact. J. Phys. B **35**, 1173 (2002)
22. Frémont, F., Bedouet, C., Tarisien, M., Adoui, L., Cassimi, A., Dubois, A., Chesnel, J.-Y., Husson, X.: Fragmentation of ions following double electron capture in slow Xe^{23+} and O^{5+} + H_2 collisions. J. Phys. B **33**, L249 (2000)
23. Olson, R.E., Feeler, C.R.: Double-electron removal from H_2 by slow, highly charged Xe^{23+} ions. J. Phys. B **34**, 1163 (2001)
24. Sobocinski, P., Rangama, J., Laurent, G., Adoui, L., Cassimi, A., Chesnel, J.-Y., Hennecart, D., Husson, X., Frémont, F.: Evidence for highly energetic fragments in O^{5+} + H_2 collisions at low impact velocities. J. Phys. B **35**, 1353 (2002)

Chapter 7
Collisions of A^{q+} and Multielectron Molecular Targets

Abstract The study of collisions between ions and multielectron molecular targets is of great importance and interest, on experimental and theoretical points of view. In this chapter, we focus on the specific case of the H_2O target. A brief discussion will be opened to larger molecules like uracil.

7.1 Interest of the Study

By studying collisions between charged particles and complex molecules involving more than two electrons, one approaches the diversity of many phenomena occurring in the universe. These complex molecules are observed for example in coma, interstellar media and in planets, especially on earth and in its atmosphere.

7.1.1 Molecules in Coma

Since the first observation of X-ray from comet Hyakutake, many observations of such rays were made on other coma. Among the mechanisms suggested for the origin of this emission (scattering of X rays from the sun, emission after collisions with electrons), the SC process between highly charged solar wind heavy ions and neutral cometary species was found to explain it in details, without any ambiguity. Again, due to the difficulty to implement quantum mechanics, the CTMC method was revealed as a powerful to explore collisions involving highly charged ions and molecules such as H_2O, CO or CO_2.

© The Author(s), under exclusive license to Springer Nature Switzerland AG 2021

F. Frémont, *Classical Treatment of Collisions Between Ions and Atoms or Molecules*,
Springer Series on Atomic, Optical, and Plasma Physics 118,
https://doi.org/10.1007/978-3-030-89428-3_7

7.1.2 *Molecular Interaction in Cells*

Ion collisions with molecules such as H_2O and constituents of matter at microscopic scale, such as uracil, thymine, or even DNA, were extensively studied experimentally and theoretically, because of their relevance in the cellular damages by ionizing radiations and their importance in ion radiotherapy. It was found for examples that low-energy electrons produced in water ionization by ion impact can lead to DNA strong breaking through the formation of autoionizing states. Therefore, the H_2O ionization and molecular fragmentation appeared as important steps in the understanding of damage in cells.

For these reasons, we focus first on the particular case of ion impacting on H_2O molecules. As in the case of small molecules reviewed in the previous chapters, total and partial cross sections will be discussed. The CTMC calculation results will be compared with experiments and/or quantum calculations when available.

In a second section, the chapter will be concluded with a discussion of preliminary studies involving big molecules such as uracil.

7.2 Collisions Involving H_2O Targets

7.2.1 *The Different Models*

Due to the complexity of the study that involves four nuclei and a large number of electrons, it was necessary to simplify the problem. The crudest approximation, if the fragmentation is not studied, is to see the molecule as an oxygen atom, with the correct binding energies for the active electrons [1]. To mimic the radial and momentum distributions of these electrons, as in the case of atomic targets, microcanonical hydrogenic distributions (h-CTMC model) or combination of microcanonical distributions (m-CTMC model) were chosen. In addition, model potentials were introduced to take into account for the evolution of the charge as seen by the active electron. Potentials of the form

$$V_{mod}(r) = V_o(r_o) + V_1(r_{H1}) + V_2(r_{H2}) \tag{7.1}$$

with

$$\begin{aligned} V_o(r_o) &= -\frac{8 - N_o}{r_o} - \frac{N_o}{r_o}\left[(1 + \alpha_o r_o)e^{-2\alpha_o r_o}\right] \\ V_H(r_H) &= -\frac{8 - N_H}{r_H} - \frac{N_H}{r_H}\left[(1 + \alpha_H r_H)e^{-2\alpha_H r_H}\right] \end{aligned} \tag{7.2}$$

were used, where r_o and r_H are the electron distances to the three target nuclei. The parameters $N_o = 7.1$, $N_H = (9 - N_o)/2$, $\alpha_o = 1.5$ and $\alpha_H = 0.665$ were found by solving approximately the Schrödinger equation to fit the energies of the valence molecular orbitals.

Very recently, this model was improved by taking into account for the dynamical response of the target [2]. The time evolution was monitored in small time steps of about 0.05 a.u. in the region where the collision happens. The time-dependent screening for the target was evaluated by making $N_o = N_o(t)$ and $N_H = N_H(t)$ on the average number of removed electrons.

7.2.2 *Total and Partial Cross Sections*

7.2.2.1 Total Cross Sections

To our knowledge, the first CTMC calculation for a H_2O target was performed in 2001 [1] for N^{7+} ion impact at energies of 2 and 4.67 keV/amu. The active electron was assumed to be bounded on *O* atom. Consequently, the molecule was considered as a pure atom, with a correct ionization energy. The experimental total SC cross section (Fig. 7.1, full squares) is in good agreement with the calculated one (full red curve). The dominant capture level $n = 5$ is well predicted by the model. However, the CTMC calculations disagree with experiment by a factor that can reach 3, for $n = 4$, 5 and 6.

The first systematic classical calculations were performed in 2007 and improved for several years, mainly for $H^+ + H_2O$ collisions and, in some extent, for He^{2+}, Li^{3+} and $C^{6+} + H_2O$ collisions [3–5]. In the case of $H^+ + H_2O$ collisions, only intermediate and high impact energies were explored, since it is well known that the CTMC model strongly deviates from other theories and experiments at low energies (typically below 20 keV).

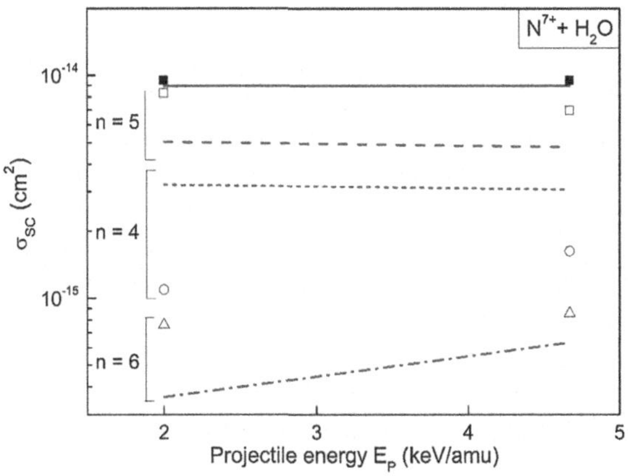

Fig. 7.1 Total and partial SC cross sections following $N^{7+} + H_2O$ collisions at energies of 2 and 4.76 keV/amu. Experiment, total and partial cross sections σ_{SC}^n ($n = 4$,5 and 6), full squares and full red curves, respectively; h-CTMC calculations, total and partial cross sections, empty objects and dashed red curves, respectively

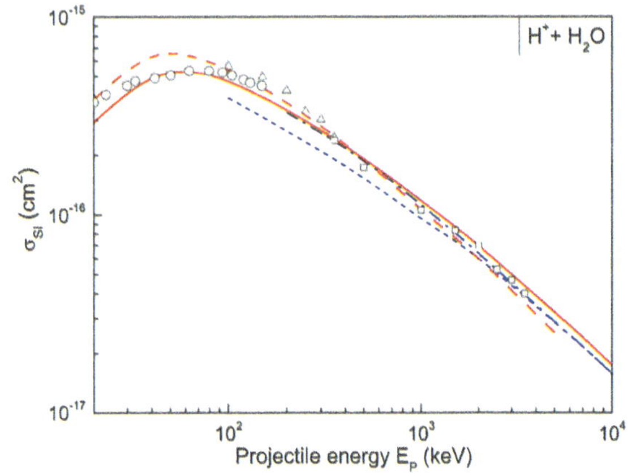

Fig. 7.2 SI cross sections following $H^+ + H_2O$ collisions, as a function of E_p [5]. Experiment, open objects; quantum calculations, dashed and dotted-dashed blue curves; CTMC calculations, dashed red curve (using Z_{eff}) and full red curve (using discretizing phase-state distributions)

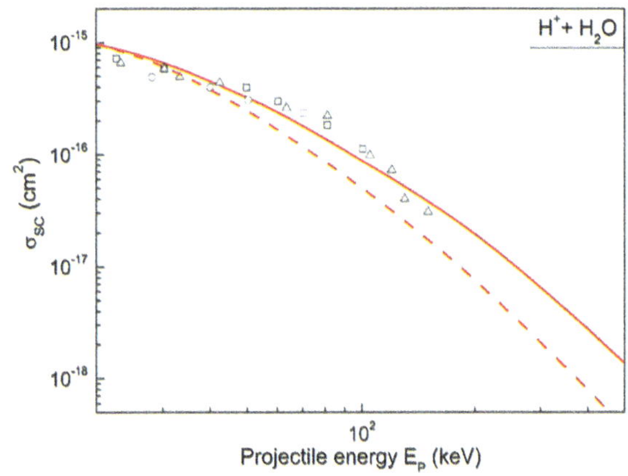

Fig. 7.3 SI cross sections following $H^+ + H_2O$ collisions, as a function of E_p [5]. Experiment, open objects; quantum calculations, dashed and dotted-dashed blue curves; CTMC calculations, dashed red curve (using Z_{eff}) and full red curve (using discretizing phase-state distributions)

Figures 7.2 and 7.3 show typical results for SI and SC, respectively, for $H^+ + H_2O$ collisions, as a function of E_p in the range from 10 to 10^4 keV [5]. In this work, the CTMC procedure, for each of the eight electrons that were considered in the collision dynamics, consisted of discretizing the phase-space distribution in terms of independent trajectories. The results (full red curves) were compared with previous CTMC results obtained from a monocentric description of the molecule in terms of an effective charge potential $-Z_{eff}/r$ (dashed red curves), quantum theories (dashed and dotted-dashed curves) and available experiments (empty objects). The classical calculations were found to reproduce very well the experimental cross sections. Using the effective potential, the SC calculations overestimate the experiments at large energies.

Very recently, two CTMC models, using static potential (sCTMC) and time-dependent mean-field calculations, were applied to He^{2+} and $Li^{3+} + H_2O$ collisions, and compared with available experiments and with two independent atom models,

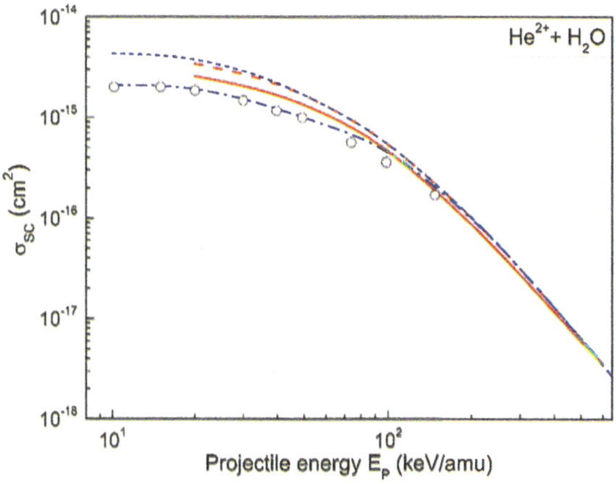

Fig. 7.4 Net capture cross sections (including SC and TI processes) following He^{2+} + H_2O collisions as a function of the projectile energy E_P. Experiment, empty circles; Calculations, short dashed and dotted-dashed blue curves, IPM-AR and IPM-PCM results, respectively; sCTMC and dCTMC results, dashed and full red curves, respectively [6]

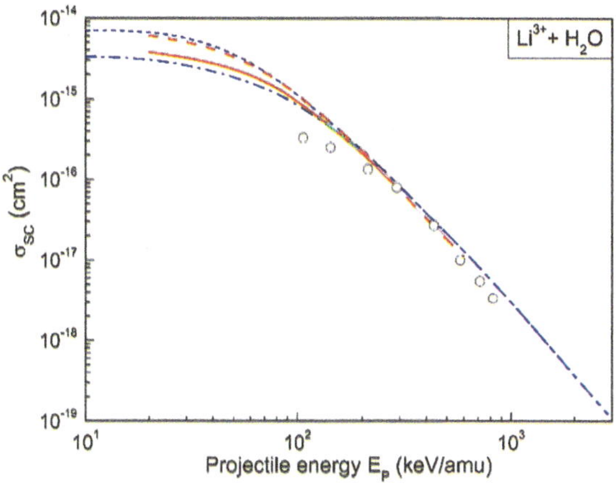

Fig. 7.5 Net capture cross sections (including SC and TI processes) following Li^{3+} + H_2O collisions as a function of the projectile energy E_P. Experiment, empty circles; Calculations, short dashed and dotted-dashed blue curves, IPM-AR and IPM-PCM results, respectively; sCTMC and dCTMC results, dashed and full red curves, respectively [6]

called IPM-AR and IPM-PCM [6]. Both models consisted of using collision information obtained for the atomic species and combining them into molecular cross sections. The IPM-AR model followed the simple Bragg additivity rule, whereas the IPM-PCM model was based on an interpretation of the atomic net cross sections[1] as geometrical areas that corresponded to capture or ionization cross section. A pixel counting method was used to measure the effective area that emerged when eliminating overlaps between cross sections from different atoms encountered by the projectile.

The results for net electron capture cross sections for He^{2+} and Li^{3+} + H_2O collisions as a function of the projectile energy E_P, ranging from 10 up to 3000 keV, is shown in Figs. 7.4 and 7.5, respectively. For He^{2+}projectiles, the dCTMC (full red

[1] The net capture (resp. ionization) cross section is the sum of the SC (resp. SI) cross sections and the TI cross section.

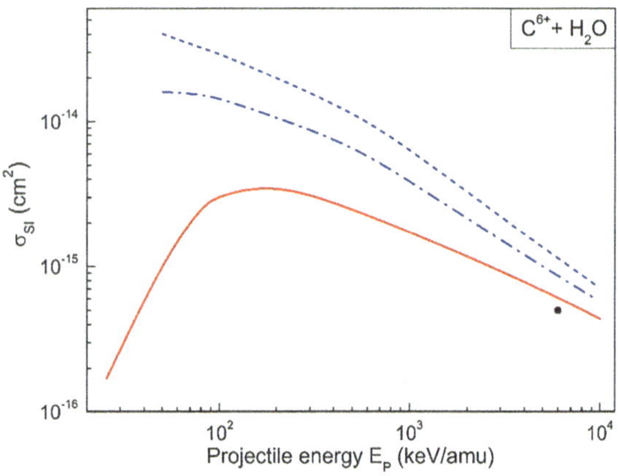

Fig. 7.6 SI cross sections following $C^{6+} + H_2O$ collisions as a function of the projectile energy E_P. Experiment, full circle; Calculations, dashed and dotted-dashed blue curves, calculations using perturbative methods; CTMC results, full red curve [4, 5]

curve) and and PCM (dotted-dashed blue curves) calculations reproduced very well the experiment (empty circles), while the AR and sCTMC models were found to be overestimated at E_P smaller than 200 keV/amu. For Li^{3+} projectiles, the same trend for the calculations was observed. However, all the calculation results were far above the experimental data at $E_P < 200$ keV/amu. From this observation, it seems clear that more experiments were needed, especially at smaller values of E_P.

Drawing from the satisfactory implementation of the CTMC model for H^+ and $He^{2+} + H_2O$ collisions, it was obvious that the present model had to be applied to $C^{6+} + H_2O$ collisions, because of the great importance of C^{6+} ions in cancer therapy [4, 5]. However, as seen in Fig. 7.6 that shows net electron capture as a function of E_P, due to the lack of experimental results (only one point at 6 meV/amu), it was impossible to give any definitive conclusion on the validity of classical calculations. In addition, the CTMC SI cross sections were found to reach a maximum value at 200 keV/amu and to decrease below this value, in contrast with results from perturbative calculations (dashed and dotted-dashed curves). The observation was identical in the case of SC (Fig. 7.7), for which, again, only one experimental result was available at ~ 3 keV/amu. At this energy, the CTMC calculations differed from experiment by a factor of about 1.5.

As mentioned previously, because the removal of one target electron during the collision is accompanied by a change of the charge as viewed by the other active electrons, a time-dependent screening method was implemented very recently to collision systems involving low-charged projectiles, such as He^{2+} or Li^{3+} [7]. Using this improvement, the dynamical response effects were taken into account. Compared with a purely static model, the SC cross sections were really improved, except for Li^{3+} projectiles. According to the authors, the problem could be in the experimental cross sections themselves. Two examples are given in Figs. 7.8. and 7.9, in the case of He^{2+} projectiles, for SC and DC, respectively. For SC, the dynamical response gave worth results compared with the static model, while DC cross sections were clearly in better agreement with the experiment, except at energies lower than 60 keV.

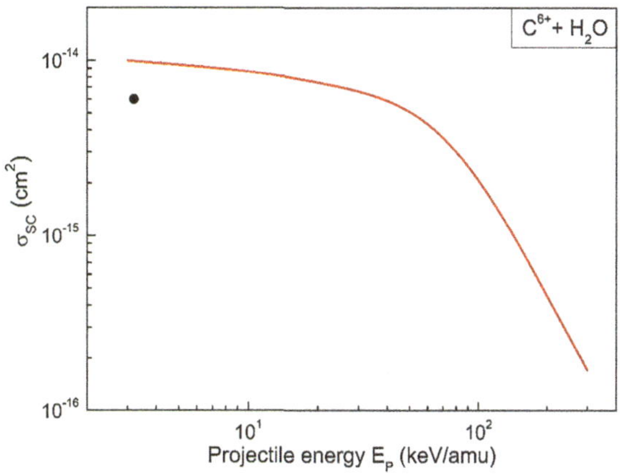

Fig. 7.7 SC cross sections following $C^{6+} + H_2O$ collisions as a function of the projectile energy E_P. Experiment, full circle; CTMC calculations, full red curve [4, 5]

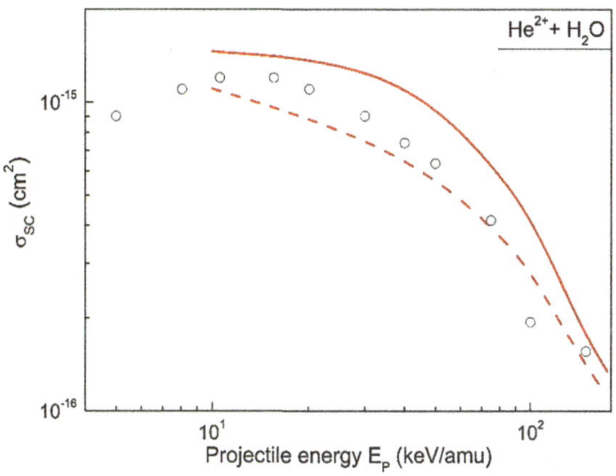

Fig. 7.8 SC cross sections following $He^{2+} + H_2O$ collisions as a function of the projectile energy E_P. Experiment, empty circles; CTMC calculations, static model, dashed red curve; dynamical response model, full red curve [7]

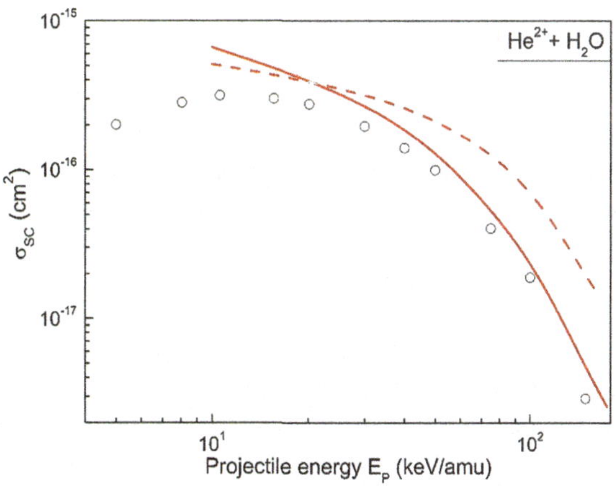

Fig. 7.9 DC cross sections following $He^{2+} + H_2O$ collisions as a function of the projectile energy E_P. Experiment, empty circles; CTMC calculations, static model, dashed red curve; dynamical response model, full red curve [7]

At this point it is necessary to comment on additions to simple models. When a model is very simple or simplified to the extreme, one can only expect to have very approximate results, which generally give an order of magnitude for the total cross sections, and this in a restricted domain of projectile energies. It is well known that improving an existing model can never replace a profound change in solving a problem. This is the case here, since even the implementation of a dynamic response improves some results at the expense of others. It is clear here that as long as the electron electron interaction is not introduced in a model, the problem will always remain intact.

7.2.2.2 Partial Cross Sections

Beyond the fundamental interest of the determination of partial cross sections following electron capture, applications in several fields have to be pointed out, such as the interactions between the solar wind and molecules constituting comets. The experimental works in this area has grown considerably at the end of the 1990s, after the discover of x-ray emission from comets. It thus became of great interest to calculate n-selective σ^n_{SC} and also l-selective σ^{nl}_{SC} cross sections, at low projectile energies. Among all the theoretical methods, the parameter-free approach was provided by the CTMC model that was able to provide at least qualitative agreement with experimental data. In order to obtain emission cross sections, cascade contributions for higher values of n were included and the nlm_l populations were multiplied by branching ratios for the relevant transitions. For the calculations, the position and momentum distributions were determined in order to match the quantum ones for the 1B1 and 3A1 orbitals. Microcanonical distributions were used, with an effective charge $Z_{eff} = 1.6$ instead of 1 in the case of hidrogenic distributions. The results for such distributions are presented in Fig. 7.10 for the two mentioned molecular orbitals [8]. Using a suitable energy distribution for the electrons, the agreement between the classical (full red curves) and quantum (dashed blue curves) distributions were reasonable.

An example of n-selective cross sections is given in Fig. 7.11 for Fe^{26+} ions impacting on H_2O at 0.01, 1 and 100 keV/amu. The n distributions were found to broaden considerably above 1 keV/amu. Similar calculations were performed for the l-selective cross sections, for given values of n. It was found obviously that small (resp. high) values of l were mainly populated at small (resp. high) energies, according to the simple formula $l = v_p b$.

To check the validity of the model, partial cross sections were calculated for O^{6+} + H_2O and compared with available experiments [8] at different projectile energies. Despite the use of model potentials to correctly describe the O^{6+} ion, especially to take into account the processes occurring at very small impact parameters, the agreement was reasonable and allowed the authors to enter into details.

The present results for the selective cross sections were used to compare theoretical and experimental line emission cross sections. Two examples are given in Fig. 7.12 for 1 keV/amu O^{8+} and Ne^{10+} + H_2O collisions. The theoretical results were normalized

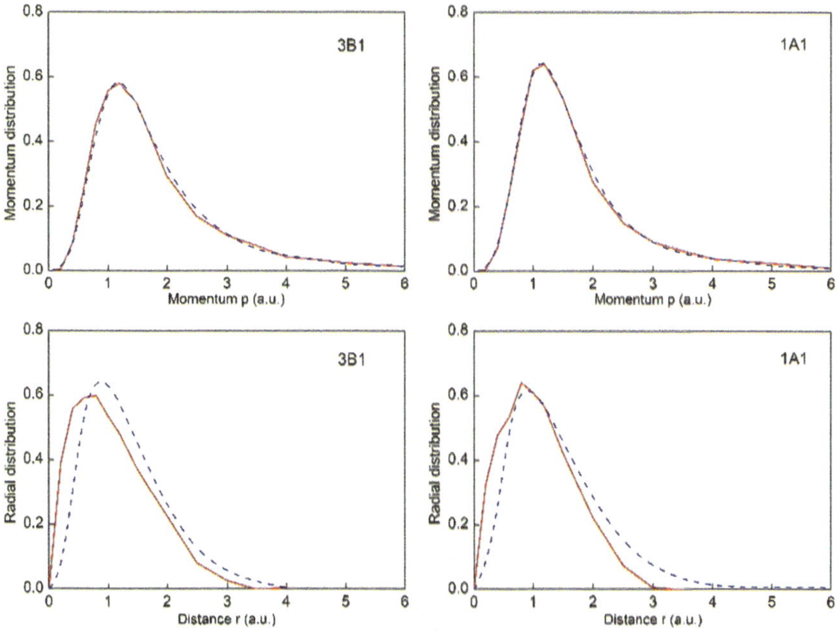

Fig. 7.10 Momentum (top) and radial (bottom) distributions of 3B1 and 1A1 orbitals of H_2O. Quantum calculations, dashed blue curves; CTMC calculations, full red curves [8]

to the experiment and convoluted in order to take into account the energy resolution. The agreement was rather good, except in the case of specific transitions (3p $\rightarrow$ 1s for example) which were a signature of DC process.

7.2.3 Multiple Electron Processes

Even though the treatment of multiple electron processes may appear as an evident task if the 8 valence electrons are considered as independent, the problem to solve is in fact a serious challenge on a quantitative point of view, since electron removal cross sections strongly depends on the binding energies of the active electrons, i.e. the effective charge of the projectile. On the other side, the inclusion of electron correlation is impossible in practice. Therefore, independent electron models are necessary to use as long as the introduction of electron correlation requires a disproportionate effort.

It is well known that single processes (i.e. processes that involve one active electron) give rise to the main contribution of the total cross section. However, multiple electron processes induce molecular fragmentation (mainly H^+ and OH^+), and the residual fragments may in turn interact with their neighbors, giving rise to secondary

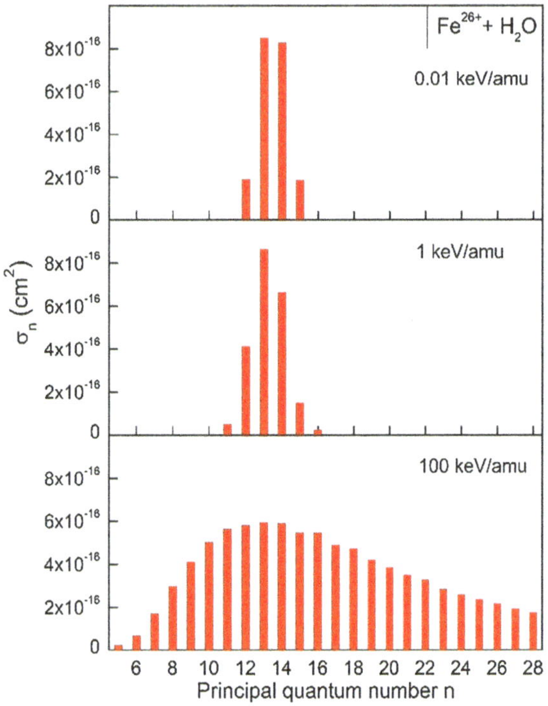

Fig. 7.11 State selective SC cross sections following $Fe^{26+} + H_2O$ collisions calculated using the CTMC method, at projectile energies of 0.01, 1 and 100 keV/amu [8]

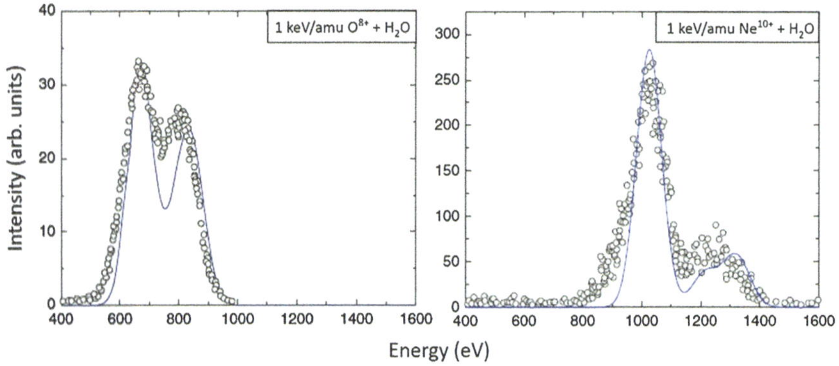

Fig. 7.12 X-ray energy emission distributions for 1 keV/amu O^{8+} and $Ne^{10+} + H_2O$ collisions [8]. Experiment, empty circles; CTMC calculations, full curves. The CTMC results were normalized to the experiment

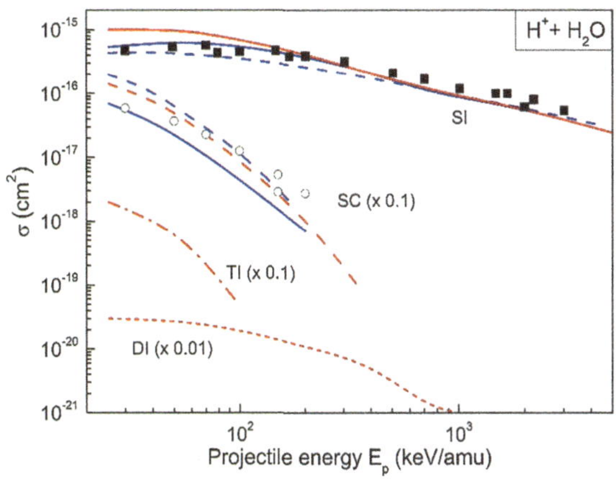

Fig. 7.13 SI, SC, TI and DI cross sections following $H^+ + H_2O$ collisions as a function of E_P. Experiment, full and empty symbols; Quantum calculations, blue curves; CTMC-COB calculations, red curves [9]

electrons or fragments. We focus here on cross sections but not on fragmentation, which will be the subject of another sub-section.

On a classical theoretical point of view, the multiple-vacancy production was given by multinomial expressions, since all the electrons were considered as independent. An original method, combining the CTMC model and COB (classical over-barrier) representation was used to describe quantitatively SI, DI, SC and DC processes following $H^+ + H_2O$ and $He^{2+} + H_2O$ collisions at energies larger than 20 keV/amu [9]. The results for calculations are presented in Figs. 7.12 and 7.13 as a function of E_P and compared with experiment when available.

The SC and SI cross sections (full red curves) were found in reasonable agreement with data at E_P larger than 100 keV/amu, but strongly overestimated the experiment at lower energies. The DC cross sections following $He^{2+} + H_2O$ collisions were smaller than the experimental one by a factor of at least 2 whatever the projectile energy. The relatively good agreement claimed by the authors allowed them to predict cross sections for larger molecules, such as cytosine and adenine, for which no experiment was available (Fig. 7.14).

An original experimental method was used to determine multiple electron removal following $A^+ + H_2O$ collisions (A = H, He and N) at an energy of about 1 meV [10]. Doubly differential fragment ion emission spectra (DDCS) in angle and energy were measured (Fig. 7.15).

The yields above 15 eV correspond to multiple ionization ($q \geq 3$). It was observed an increase of the multiple ionization process when increasing the mass of the nucleus, due to the weaker screening of the projectile nucleus at small impact parameters.

From these spectra, absolute cross sections for the individual fragmentation channels were determined, with the knowledge of the majority of the kinetic energy release (KER) of the fragments. Finally, from these cross sections, the multiple ionization cross sections were deduced, and compared with CDW-EIS and CTMC calculations, for ionization degrees between 1 and 5. Even though a relatively good agreement was found for small ionization degrees, discrepancies were found for $q = 4$ and 5.

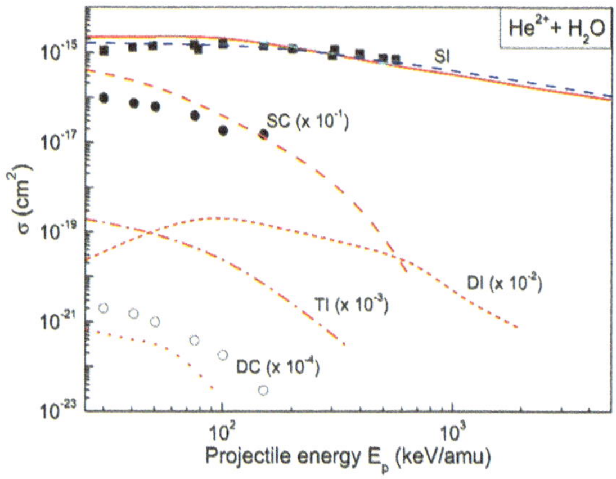

Fig. 7.14 SI, SC, TI, DI and DC cross sections following $He^{2+} + H_2O$ collisions as a function of E_P. Experiment, full and empty symbols; Quantum calculations, blue curves; CTMC-COB calculations, red curves [9]

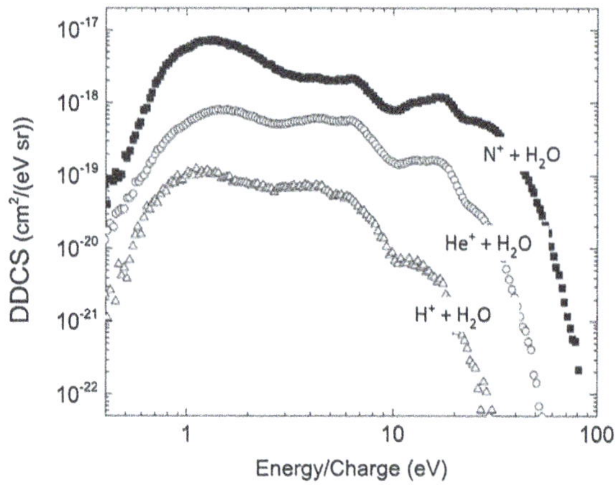

Fig. 7.15 Typical energy spectra of emitted fragments following $A^+ + H_2O$ collisions (A = H, He and N) at an energy of about 1 meV [10]. H^+, open triangles; He^+, open circles; N^+, full squares

The measured cross sections decreased faster than the calculated one. This tendency was attributed by the authors to the limitation of the independent particle model.

7.2.4 *Differential Cross Sections*

As usual, the determination of SDCS and DDCS constitutes a stringent test of the methods employed, and provides data for several codes, especially those that simulate ion tracks in biological materials. The CTMC method was applied for example to determine SDCS $d\sigma/dE_e$ following $A^+ + H_2O$ collisions at intermediate and large E_P [11]. An example of SDCS is reported in Fig. 7.16, for $E_P = 100$ and 150 keV. The CTMC results (full red curves) matched very well the experiment (open circles). The FBA calculations (dashed blue curves) overestimated the cross sections at low

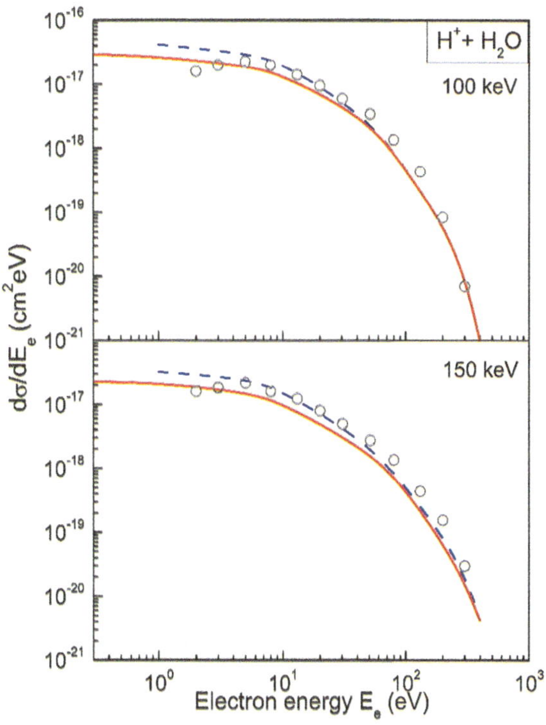

Fig. 7.16 SDCS following $H^+ + H_2O$ collisions at projectile energies of 100 and 150 keV [11]. Experiment, empty circles; Quantum calculations using FBA, dashed blue curves; CTMC calculations, full red curves

electron energies. When entering into the details, it was observed that the calculations could not reproduce the maximum of the SDCS seen experimentally at $E_e \sim 4$ eV. Despite this small deviation, it was clear that the CTMC method was able to provide accurate SDCS for E_P larger than 20 keV.

Similar studies were performed a few years later for the same collision system, as well as for He^+ and $N^+ + H_2O$ collisions [12]. DDCS in energy and in emission angle were measured, and the emitted electron energy range was extended to 3000 eV in order to enlighten the binary encounter structures, and compared with CDW-EIS and CTMC calculations.

For bare projectiles, potential energies V_{mod}, as expressed in (7.1) and (7.2) were used. For structured projectiles, especially N^+, which has 6 bounded electrons, the Green-Sellin-Zachor (GSZ) potential forms were applie for the determination of electron projectile and target projectile interactions. A similar approach was used for H_2O, in which 8 electrons belong to O, and 2 electrons on H. A better approximation consisted of a small change in these numbers, since the H atoms are not completely screened, i.e. a small part of the electron cloud is distributed on the O atom. As a result, $8 + \Delta N$ electrons were supposed to orbit around O, and $2 - \Delta N/2$ electrons on H. The best value for ΔN was found to be 0.1.

Two examples of DDCS are shown in Fig. 7.17 for $H^+ + H_2O$ (top of the figure) and $N^+ + H_2O$ (bottom of the figure) collisions at $E_P = 1$ meV, at different electron

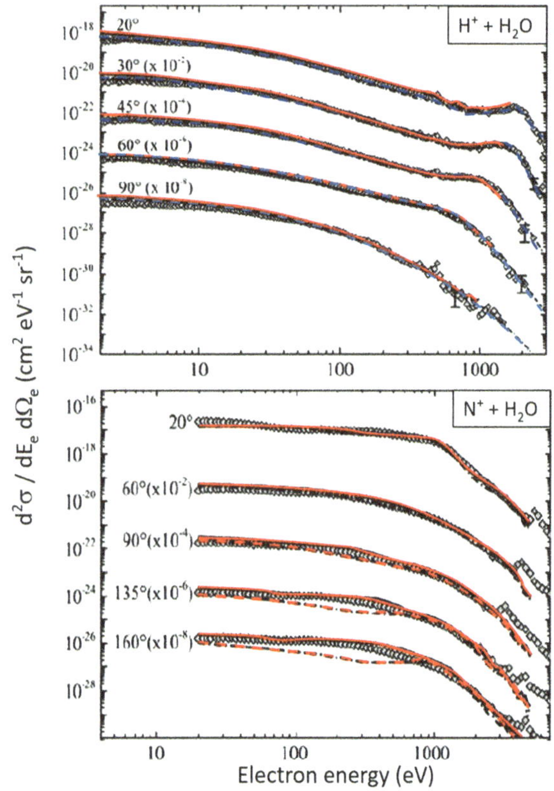

Fig. 7.17 DDCS following $H^+ + H_2O$ (top) and $N^+ + H_2O$ (bottom) collisions at 1 meV, as a function of the emitted electron energy, for different detection angles [12]. Experiment, empty diamonds; CDW-EIS calculations, dashed blues curves; CTMC calculations without projectile ionization, dashed red curves; CTMC calculations including projectile ionization, full red curves

emission angles. The CTMC results (full red curves) were in very good agreement in shape and in absolute values with experimental spectra (empty diamonds) and CDW-EIS calculations (dashed blue curves). A systematic deviation of calculations was observed below 10 eV, and was partially attributed to a systematic error in the measurement. The agreement between experiment and classical results were also very good in the case of N^+ projectiles, with the condition that, in addition to target ionization (dashed red curves), the projectile ionization was included (full red curves). It was noted discrepancies when using quantum calculations (not shown here).

More recently, similar theoretical studies were performed using fully stripped C^{6+}, $O^{8+}$and Si^{13+} ion projectiles at velocity of about 10 a.u. [13]. DDCS and SDCS for target ionization were calculated and compared to recent experimental and theoretical data. Static and time-dependent screening approaches were implemented. The last approach took into account the removal of electrons during the collision to update the charge state of the target at each time step. DDCS were calculated for two values (5 and 200 eV) of the emitted electron energy E_e. The results are presented in Fig. 7.18 for 200 eV, for detection angles varying between 30° and 150°. The experimental results (full circles) were compared with two CDW-EIS models (blue curves), and the static (full red curves) and dynamic (dashed red curves) CTMC calculations.

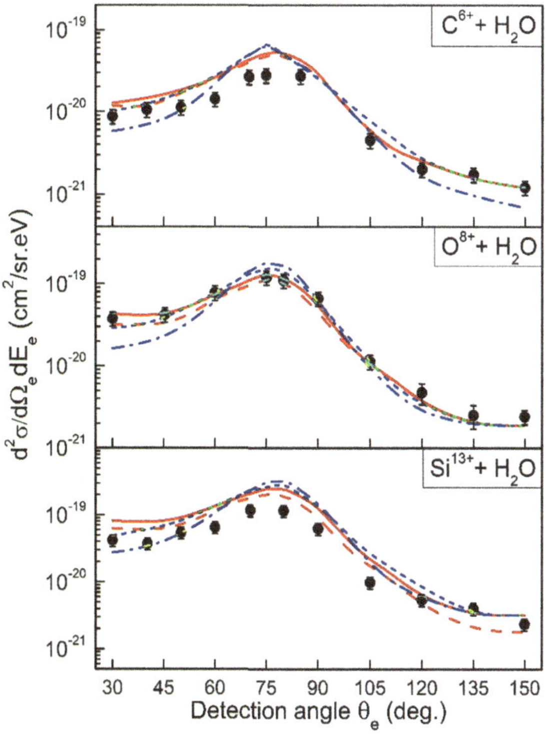

Fig. 7.18 DDCS for electron emission following A^{Z+} (A = C, O and Si) + H_2O collisions at high impact velocities, for an electron emission energy of 200 eV [13]. Experiment, full circles; CDW-EIS calculations, dotted-dashed and short dashed blue curves; CTMC calculations, using static and time-dependent screening, full and dashed red curves, respectively

At 200 eV emission energy, the ionization mechanism can be considered classical, and therefore the CTMC method is expected to work properly. As seen in Fig. 7.18, the CTMC DDCS were consistent with the experiment, especially for O^{8+} projectiles for which an almost perfect quantitative agreement was observed in shape and in maximum. For the other projectiles, the CTMC results were overestimated. In contrast, in all cases, the quantum calculations overestimated the measurements, and this overestimation was much greater for C^{6+} and Si^{13+}. Similar trends were also found for emission energies of 5 eV. The authors attributed the differences for these two projectiles to a problem with the normalization factor in the experiment.

It was noted that the calculated total DDCS and SDCS using the CTMC method were not affected by the nature of the screening. However, when examining the relative importance of multifold contributions, differences appeared between static and dynamical screening. These differences are summarized in Fig. 7.19, for single up to fivefold ionization. For a degree ionization q equal to 1 and 2, both models gave similar results. Huge differences were observed for larger degrees of ionization. The cross section is quite insensitive to q, whereas a strong decrease of the cross section was found when using the dynamic screening. Unfortunately, no experimental data were available to confirm or not the two approaches.

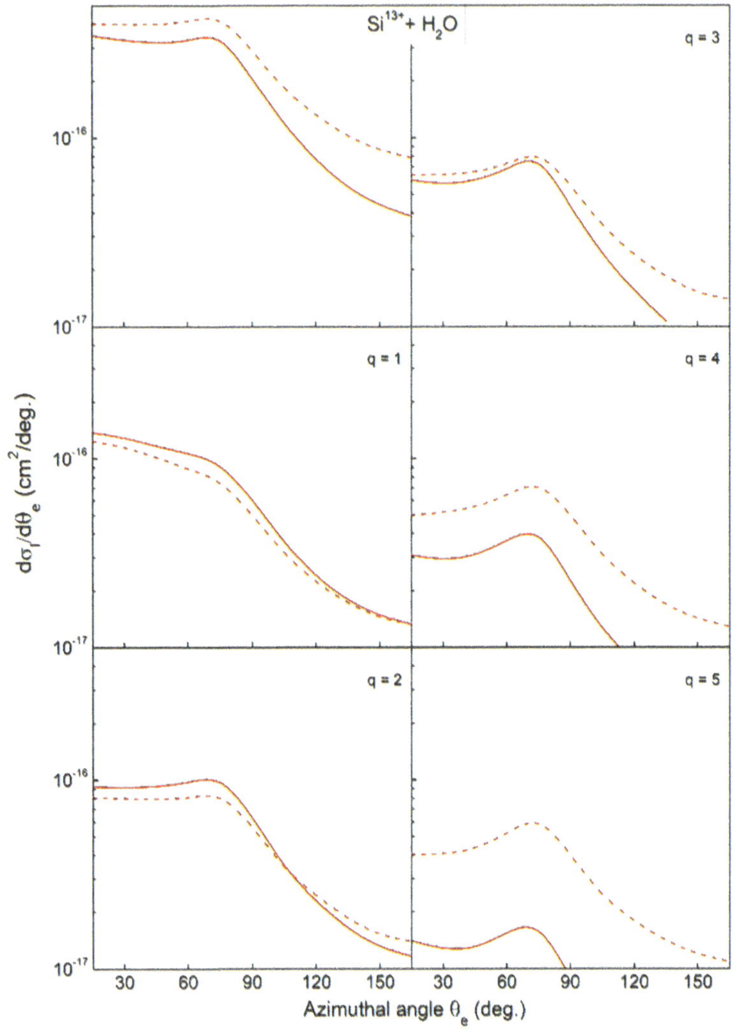

Fig. 7.19 Calculated CTMC SDCS for multifold ionization ($q = 1 - 5$) and total SDCS following $Si^{13+} + H_2O$ collisions [13]. Static screening, dashed curves, dynamic time-dependent screening, full curves

7.2.5 *Fragmentation of the H_2O Molecule*

In the same way as the H_2 molecule, it is well known that the H_2O, even though it mainly remains H_2O^+ after single electron processes, may also fragment, especially when more than one electron is removed. As mentioned in the introduction, fragmentation reactions are of great importance in the radiation damage of biological species because the secondary ions can interact with DNA. The main dissociation channels for H_2O^+ write

$$
\begin{aligned}
H_2O^+ &\rightarrow OH^+ + H \quad \text{evaporation} \\
H_2O^+ &\rightarrow OH + H^+ \quad \text{fission} \\
H_2O^+ &\rightarrow H + H + O^+ \quad \text{breakup}
\end{aligned}
\tag{7.3}
$$

In order to determine $\sigma_{SI}^{A^+}$ and $\sigma_{SC}^{A^+}$ for the production of a A^+ fragment following SI and SC processes, the SI and SC cross sections were weighted by different branching ratios for fragmentation [14, 15].

As an example, total cross sections for formation of H_2O^+, OH^+, H^+ and O^+ following SI and SC in $H^+ + H_2O$ collisions are presented in Figs. 7.19 and 7.20 as a function of the projectile energy. CTMC calculations using branching ratios from Tan [14] were compared with experimental data [5]. For both processes, the agreement was found quantitatively good, except at energies smaller than 300 keV for SI. For SC process, a maximum appears in the experiment at about 35 keV which is not reproduced by the calculations. Despite these details, the agreement is at least as good as the agreement using quantum calculations [16] (Fig. 7.21).

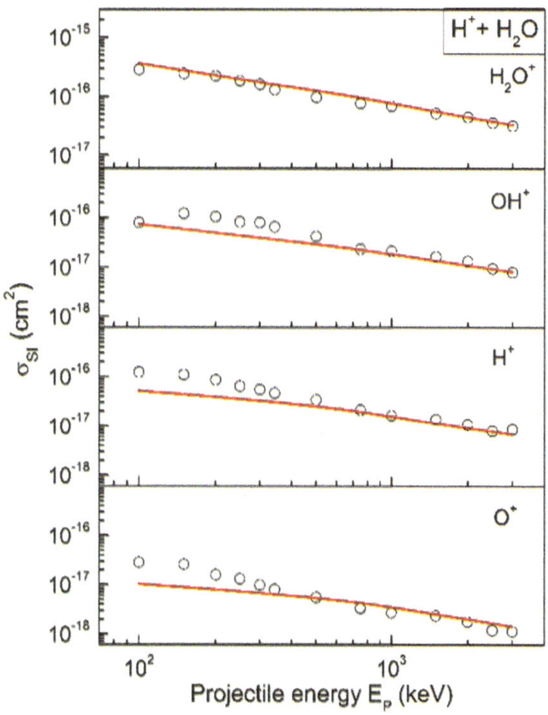

Fig. 7.20 Cross section for production of H_2O^+, OH^+, H^+ and O^+ ions following SI in $H^+ + H_2O$ collisions, as a function of the projectile energy [5]. Experiment, empty circles; CTMC calculations, full red curves

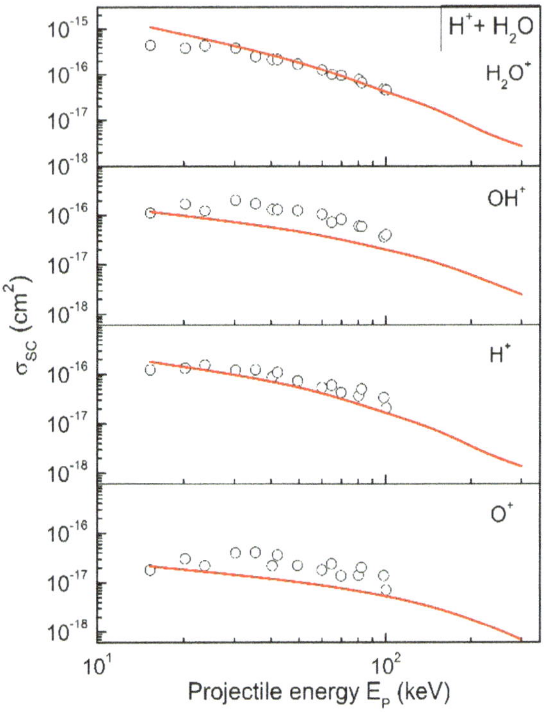

Fig. 7.21 Cross section for production of H_2O^+, OH^+, H^+ and O^+ ions following SC in $H^+ + H_2O$ collisions, as a function of the projectile energy [5]. Experiment, empty circles; CTMC calculations, full red curves

7.2.6 Introduction of Quantum Effects

The introduction of quantum effects in the the calculations was made in 1997 [17] to study proton stopping on atomic targets and H_2O molecules. Since electron correlation gives rise to unstabilities of the molecule, it was neglected in the calculations. In addition, the atomic centers were fixed during the collision, and the energy of the initial electron distribution was minimized by adjusting the size of the Heisenberg cores in the expression of the phenomenological potentials V_H and V_P, in order to achieve the correct total binding energy of ~76.5 a.u. The results of the proton stopping power calculatedusing the CTMC method are shown in Fig. 7.22 (full red curve) in the energy range 10 keV–100 meV, and were compared with a reference curve (dashed blue curve) calculated from experimental datat for O and H. The fairly good agreement was a clear indication of the predominance of proton-electron interactions in this process.

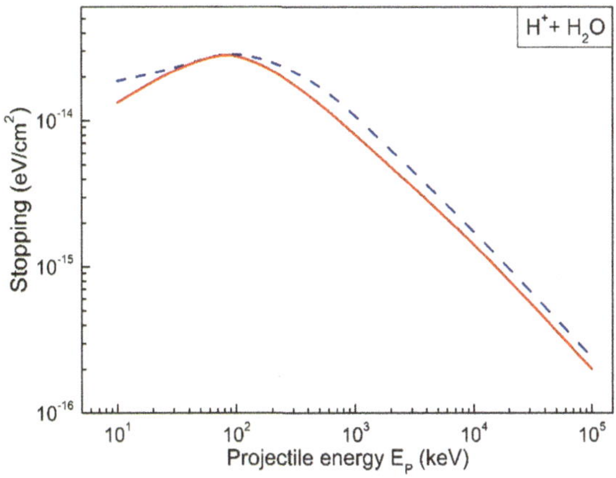

Fig. 7.22 Proton stopping following $H^+ + H_2O$ collisions as a function of the projectile energy [17]. CTMC calculations, full red curve; reference curve derived from experiment, dashed blue curve

7.3 Collisions Involving Biological Molecules

Concerning large molecules such DNA and RNA component targets, the existing experimental and theoretical studies remain rare. Experimental works were dealing essentially with RNA uracil and bromuracil bases impacted by low energy electrons. Collisions of such species with multiply charged ions are even scarcer. In 2009, collisions of H^+, He^{2+} and C^{6+} with DNA nucleobases and uracil were investigated theoretically in order to determine SC and SI cross sections at E_p ranging from 10 keV up to 10^4 keV/amu [18]. The CTMC method combined with the COB (classical over barrier) model was used. To simplify, the target position was fixed in the laboratory frame, whereas the active electrons were, in a first step, considered as virtual particles and taken into account only if ionization or capture processes occurred. The initial orientation of the molecules was not taken into account. If an electron was effectively ejected, a new potential, according to the COB model, was evaluated. At the beginning of the collision, only the projectile and the target existed. To evaluate the possibility for an electron to be ejected, two simultaneous conditions had to be fulfilled:

- The maximum of the potential energy had to be lower than the electron binding energy.
- The ratio $\Delta t/t_e$, where Δt and t_e were the integration time step and the classical orbital period of the target electron, respectively, had to be larger than a random number chosen in the (0,1) interval.

The calculations required only the knowledge of the electron binding energies. In Fig. 7.23, the present calculations for the collision system H^+ + uracil (full red curve) were compared to the only existing FBA calculations (dashed blue curve) for E_p in the range 100–10,000 keV. Strong and systematic discrepancies in magnitude and in shape were observed, but could not be explained and were surprising, since recent FBA calculations on water ionization following H^+ impact were in good agreement with CTMC-COB predictions. Moreover, the present CTMC calculations matched

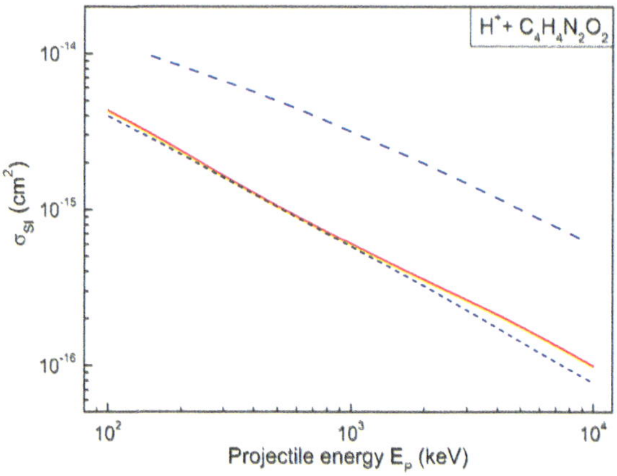

Fig. 7.23 Calculated SI cross sections following H^++ uracil collisions as a function of E_p [18]. CTMC-COB calculations, full red curve; FBA calculations, dashed blue curve; empirical formula, short dashed blue curve

cross sections obtained with a simple formula (short dashed blue curve) determined with a simple formula obtained by using the Rutherford formula differential in energy and considering the well known $ln E_p/E_p$ dependence.

In 2015, the electron emission from the uracil molecule induced by fast proton impact was investigated using the CTMC method [19]. The independent-particle model was applied, and only one electron was considered as active. The interactions with the molecular target were described by a multicenter potential built from screened atomic potentials. The goal of this work was to determine DDCS ans SDCS as well as total ionization cross sections, and to compare with FBA, CDW-EIS calculations, CTMC-COB model and available experiments.

In the present classical model, modified Green-Sellin-Zachor potentials were used, introducing a change ΔN of the electron number from N to $N + \Delta N$. The total potential was written as a sum of individual potentials for H, C, N and O, which were considered as fixed. In the initial conditions, 29 molecular orbitals were considered, depending on the electron binding energies.

As an example, a contour map of the initial electron position is shown in Fig. 7.24 (left side) for a loosely bound electron (0.581 a.u.), and the trajectory of this electron is reported on the right of Fig. 7.24, to clearly show that the trajectory reflects the initial conditions.

Figure 7.25 shows DDCS for 1 meV protons as a function of the emitted electron energy E_e, for detection angles of 15, 90 and 165°. The CTMC calculations (full red curves) were in good agreement with experiment (empty circles), except at energies smaller than 10 eV, and at energies larger than 200 eV at 165°. Part of the ionization at large energies was found to originate from inner molecular orbitals, but this contribution, which explained the small bump observed at about 500 eV at 15° is not large enough at large angles. The CTMC calculations were found to match the FBA results (dashed blue curves), whereas CDW-EIS calculations (short dashed

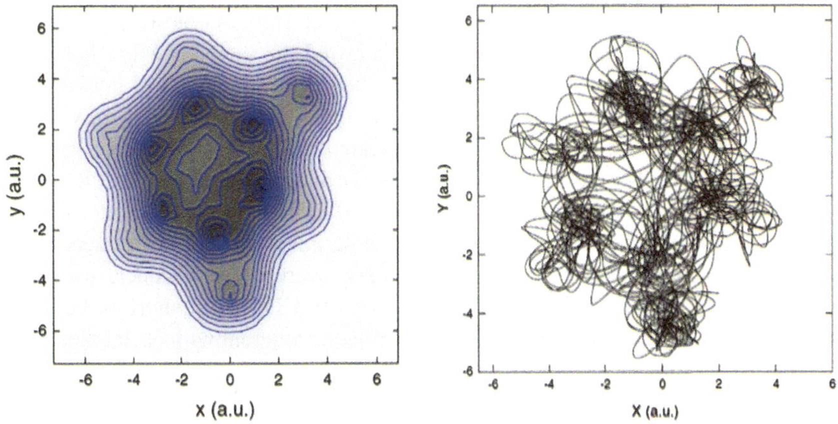

Fig. 7.24 Contour map (left side) of the initial electron positionin uracil. The binding energy of the electron is −0.581 a.u. x and y are coordinates of the projection of the position vectors into the molecule plane. On the right side, an example of trajectory in the xOy plane is shown [19]

Fig. 7.25 DDCS for SI following 1 meV protons colliding on uracil as a function of the emitted electron energy, at detection angles of 15, 90 and 165° [19]. Experiment, empty circles; CTMC calculations, full red curve; FBA results, dashed blue curve; CDW-EIS results, short dashed blue curve

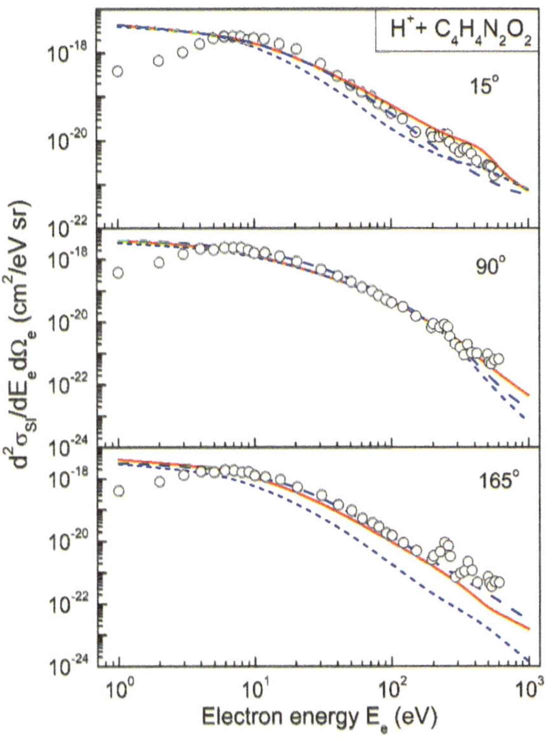

blue curves) deviated from experiment, especially at 165°. As far as the projectile energy decreased, calculated cross sections followed the same behavior than those calculated with quantum methods. However, the difference observed between the calculation and the experiment was much larger than that observed at 1 meV.

The disagreement between experiment and calculations is even poorer at 100 keV, as seen in Fig. 7.26. The calculations differ from the data (empty circles) by two orders of magnitude at E_e smaller than 100 eV, whereas CTMC calculations (dotted dashed red curve) performed at the same time than experiment were in better agreement with the experiment, even though they were clearly overestimated. Conclusions were impossible to make since FBA (dashed blue curve), CDW-EIS (short dashed blue curve) and CTMC calculations (full red curve) were coherent with each other.

After integration of DDCS on energy and angle, the comparison of the CTMC results for total SI cross sections with CDW-EIS, FBA, CTMC-COB models and experiment (Fig. 7.27) strong divergences of the models with decreasing the impact

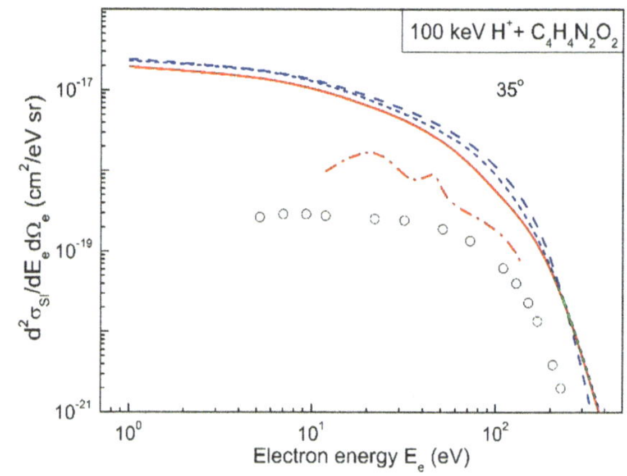

Fig. 7.26 DDCS for SI following 100 keV protons colliding on uracil as a function of the emitted electron energy, at detection angles of 15, 90 and 165° [19]. Experiment, empty circles; CTMC calculations, full red curve; FBA results, dashed blue curve; CDW-EIS results, short dashed blue curve; previous CTMC calculations, dotted dashed red curve

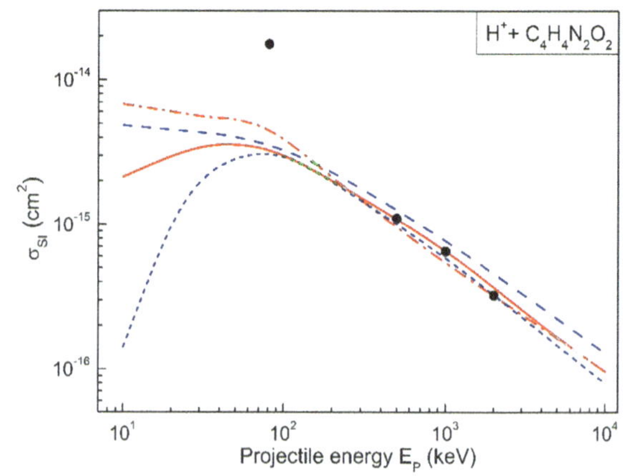

Fig. 7.27 Total SI cross section following collisions of protons on uracil as a function of the projectile energy [19]. Experiment, full circles; CTMC calculations, full red curve, CTMC-COB calculation, dotted dashed red curve; FBA results, dashed blue curve; CDW-EIS results, short dashed blue curve

energy. This called, according to the author, the attention for the necessity of experiments below 100 keV to efficiently check the validity of the models.

7.4 Track Structure Simulations

7.4.1 Principle of the Method

Beyond the fundamental aspect, the precise knowledge of differential cross sections in energy and in angle has a practical interest. Several areas are of great demand of quantitative issues, such as the nature of the damage formed by ionizing radiation for DNA repair studies, or exposure to space radiation which represents another situation where human beings can be irradiated by heavy ions. The latter example is made particularly complex by the fact that particles originating from universe are a mixed field of ~87% highly energetic protons, ~12% He fully stripped ions and ~1% of heavier ions.

To follow the damages in matter, many Monte Carlo codes were developed and extensively used [20–22]. The goal was to determine, from elastic and inelastic process cross sections calculating using quantum or classical mechanics, many useful quantities that are important to know, such as the total inelastic free path (IFP), the radiation-induced energy deposits at the nanoscale level in the medium of interest, yields of DNA fragments of different sizes, or track-structure representations. Note that these quantities can be considered as classical in the sense that, even though probabilities are calculated, the trajectories of the projectile and the various targets induced during the passage of the projectile are followed.

7.4.2 A Few Results

Due to the complexity of the problem, only a few simple results are given here, without entering into the details of the calculations. Figure 7.28 shows MC simulation data for single-strand break (SSD) and double-strand brake (DSB) yield per cell for electrons (full diamonds), protons (full circles) and He^{2+} ions (full squares), as a function of the parameter $\left(Z_{eff}/\beta\right)^2$, where Z_{eff} is the effective charge of the projectile, and β is defined as the ratio between the projectile velocity and the velocity of light in vacuum. The thick red curves originated from a function that could fit the data, with the following form

$$f\left(Z_{eff}/\beta\right) = a + \frac{(b-a)\left(Z_{eff}/\beta\right)^2}{\left(Z_{eff}/\beta\right)^2 + c} \tag{7.4}$$

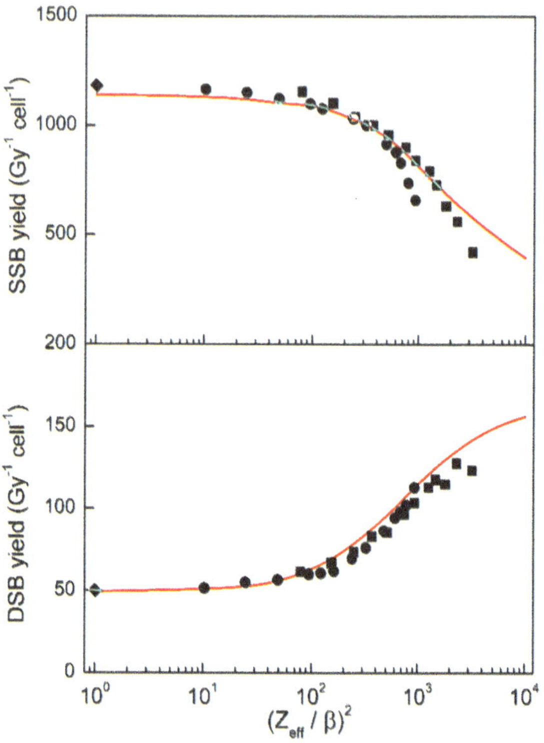

Fig. 7.28 Results of SSB and DSB yields per cell following collisions between electrons (full diamonds), protons (full circles) and helium ions (full squares), as a function of the parameter $(Z_{eff}/\beta)^2$ [20]. The full red curves were derived from relation 7.4

where *a*, *b* and *c* were adjustable parameters.

From formula (7.4) parameters for MC simulations were derived, and SSB and DSB yields could be calculated as a function of the projectile energy. The results are presented in Fig. 7.29 for *He* ion projectiles (full red curves) and compared with previous theoretical results (full circles and squares). The agreement was found to be reasonable whatever the projectile energy.

To examine the effects on damage complexity of the direct and indirect DNA damage mechanisms, the MC algorithm was modified to mimic reductions in the amount of strand breaks and base damages associated with exposure to DSMO (dimethyl sulfoxide) which is a free radical scavenger. It was assumed that the number of strand breaks and base damages decreased as the DSMO concentration increased, by multiplication using a dimensionless function that depended mainly onto the scavenger concentration and the fraction of strand breaks and base damages that were not scavengeable. It was shown that MC simulations could very reasonably approximate the experimental data for different DSMO concentrations. Finally, the analysis suggested that the indirect effect of ionizing radiation contributed to the formation of more complex forms of DNA strand breaks.

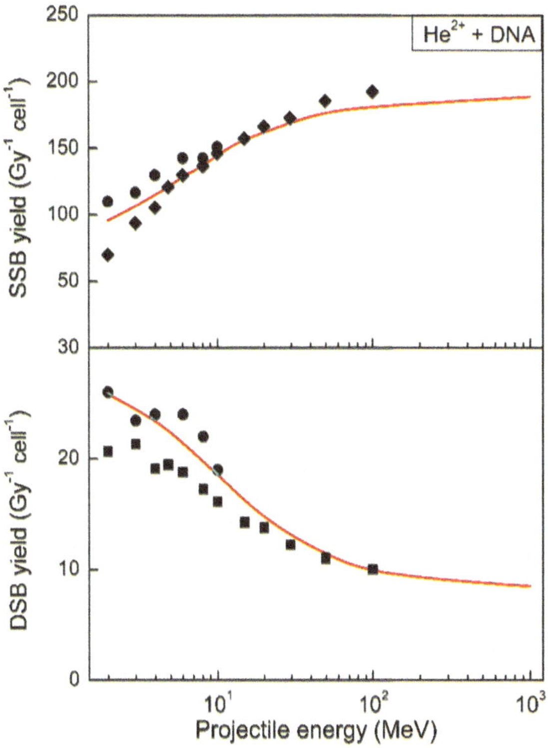

Fig. 7.29 Production of SSB and DSB following He^{2+} ions colliding on DNA [20], as a function of the projectile energy. Calculations derived from the semi empirical formula (7.4), full red curves; Other calculations, full circles and squares

References

1. Hasan, A.A., Eissa, F., Ali, R., Schultz, D.R., Stancil, P.C.: State-selective charge transfer studies relevant to solar wind-comet interactions. The Astrophys. J. **560**, L201 (2001)
2. Jorge, A., Horbatsch, M., Kirchner, T.: Multicharged-ion-water-molecule collisions in a classical-trajectory time-dependent mean-field theory. Phys. Rev. A **102**, 012808 (2020)
3. Errea, L.F., Illescas, C., Méndez, L., Pons, B., Rabadan, I., Riera, A.: Classical calculations of ionization and electron capture total cross sections in $H^+ + H_2O$ collisions. Phys. Rev. A **76**, 040701(R) (2007)
4. Méndez, L., Errea, L.F., Illescas, C., Rabadan, I., Pons, B., Riera, A.: Calculation of total cross sections for ionization and charge transfer in collisions of multicharged ions with water molecules. In: Tökesi, K., Sulik, B. (eds.) Rad. Dam. in Bio. Syst., Proceedings of the 5th International Conference (RADAM) (2008)
5. Illescas, C., Errea, L.F., Méndez, L., Pons, B., Rabadan, I., Riera, A.: Classical treatment of ion- H_2O collisions with a three-center model potential. Phys. Rev. A **83**, 052704 (2011)
6. Lüdde, H.J., Jorge, A., Horbatsch, M., Kirchner, T.: Net electron capture in collisions of multiply charged projectiles with biologically relevant molecules. Atoms **8**, 59 (2020)
7. Jorge, A., Horbatsch, M., Kirchner, T.: Multicharged-ion–water-molecule collisions in a classical-trajectory time-dependent mean-field theory. Phys. Rev. A **102**, 012808 (2020)
8. Otranto, S., Olson, R.E.: Charge exchange and x-ray emission cross sections for multiply charged ions colliding with H_2O. Phys. Rev. A **77**, 022709(R) (2008)

9. Abbas, I., Champion, C., Zarour, B., Lasri, B., Hansen, J.: Single and multiple cross sections for ionizing processes of biological molecules by protons and α-particle impact: a classical Monte Carlo approach. Phys. Med. Biol. **53**, N41 (2008)
10. Kovács, S.T.S., Herczku, P., Yuhász, Z., Sarkadi, L., Gulyás, L., Sulik, B.: Dissociative ionization of the H_2O molecule induced by medium-energy singly charged particles. Phys. Rev. A **96**, 032704 (2017)
11. Errea, L.F., Illescas, C., Méndez, L., Riera, I.: Ionization of water molecules by proton impact: Two nonperturbative studies of the electron emoission spectra, Phys. Rev. A **87**, 032709 (2013)
12. Kovács, S.T.S., Herczku, P., Yuhász, Z., Sarkadi, L., Gulyás, L., Sulik, B.: Ionization of small molecules induced by H^+, He^+ and N^+ projectiles: comparison of experiment with quantum and classical calculations. Phys. Rev. A **94**, 012704 (2016)
13. Jorge, A., Horbatsch, M., Illescas, C., Kirchner, T.: Classcal-trajectory Monte Carlo calculations of differential electron-emission cross sections in fast heavy-ion collisions with water molecules, Phys. Rev. A **99**, 062701 (2019)
14. Tan, K.H., Brian, C.E., Van der Leeuw, P.E., Van der Wiel, M.J.: Absolute oscillator strengths (10–60 eV) for the photoabsorption, phoionisation and fragmentation of H_2O. Chem. Phys. **29**, 299 (1978)
15. Olivera, G.H., Caraby, C., Jardin, P., Cassimi, A., Adoui, L., Gervais, B.: Multiple ionization of water radiolysis. Phys. Med. Biol. **43**, 2347 (1998)
16. Murakami, M., Kirchner, T., Horbatsch, M., Lüdde, H.J.: Single and multiple electron removal processes in proton-water-molecule collisions. Phys. Rev. A **85**, 052704 (2012)
17. Becks, W.A., Wilets, L.: Semiclassical description of proton stopping by atomic and molecular targets. Phys. Rev. A **55**, 2821 (1997)
18. Lekadir, H., Abbas, I., Champion, C., Fojon, O., Rivarola, R.D., Hanssen, J.: Single-electron-loss cross sections of DNA and RNA bases impacted by energetic multicharged ions: A classical Monte Carlo approximation, Phys. Rev. A **79**, 062710 (2009)
19. Sarkadi, L.: Classical treatment of the electron emission from collisions of uracil molecules with fast protons. Phys. Rev. A **92**, 062704 (2015)
20. Semenenko, V.A., Stewart, R.D.: Fast Monte Carlo simulations of DNA damage formed by electrons and light ions. Phys. Med. Biol. **51**, 1693 (2006)
21. Ballarani, F., Alloni, D., Facoetti, A., Ottolenghi A.: Heavy-ion effects from track-structure to DNA and chromosome damage, New J. Phys. **10**, 075008 (2008)
22. Alcocer-Avila, M.E., Quinto, M.A., Monti, J.M., Rivarola, R.D., Champion, C.: Proton transport modeling in a realistic biological environment by using TILDA-V. Sci. Rep. **9**, 14030 (2019)

Chapter 8
Conclusions and Perspectives

8.1 Conclusions

Quantum mechanics, which takes its origin from quantum physics and principles, has been considered since the beginning of the twentieth century as the only tool that could be able to treat interactions between particles, in order to obtain essential qualitative and quantitative insights to the knowledge of elastic and inelastic processes in atomic collisions. Classical mechanics, which was far to be put aside, was used from the beginning of the same years. Based on the resolution of time-dependent Hamilton equations, it associates trajectory calculations (determinism) and probabilities (indeterminism) to calculate the same quantities as those calculated with quantum mechanics.

We have reviewed the classical methods used to date and, especially, the CTMC method. It turned out that this method presents many advantages, compared to quantum mechanics:

- It simulates collision taking into account multiple parameters, such as the target and projectile positions, their energy, the number of active electrons, the impact parameter.
- It allows to calculate any quantity, whether it is total, partial or differential absolute cross sections, or energy or angular distributions.
- It is able now to integrate the electron correlation all along the collision, as well as interferences observed in several distributions.
- It describes partially some pure quantum phenomena, such as the Auger effect, or quantum numbers n, l and m_l.

The CTMC method is used for collisions assumed to be simple, such as $H^+ + H$, and, at the opposite, for complex collisions, involving several active electrons and nuclei. It appears that, on the one hand, these classical methods, used also in nuclear physics, have recognized bases and sufficiently solid to refine calculations and to

© The Author(s), under exclusive license to Springer Nature Switzerland AG 2021

F. Frémont, *Classical Treatment of Collisions Between Ions and Atoms or Molecules*, Springer Series on Atomic, Optical, and Plasma Physics 118,
https://doi.org/10.1007/978-3-030-89428-3_8

carry out other studies, and, on the other side, one can say today that quantum and classical methods are not opposed but complementary.

It seams now obvious that classical calculations are so promising that they are able to be conducted on to ways. First, they can be extended to more complex systems than those studied so far and, second, refined analysis on simple systems where quantum calculations are not in agreement with experiment can be performed. Rather than focusing on complex systems, we give in the following some perspectives concerning systems involving small targets, such as H, He and H_2.

8.2 Perspectives

8.2.1 $H^+ + H$ Collisions

As shown in Chap. 3, the CTMC method is well adapted, in the case of $H^+ + H$ collisions, for the determination of total capture cross sections at energies larger than 20 keV. At smaller energies, whatever the potentials used, the calculated cross sections are underestimated. It is thus clear that these models have to be improved or changed for new models.

As in any problem to solve, there are at least two ways to proceed with when disagreements occur with experiments or other theories. Either the model that was used previously is improved to take into account the specificity of the problem, or the model is no longer used and replaced by a new model. In the present case, we only invoke the possibility to improve the model.

In the CTMC method, two series of parameters have to be taken into account: the initial radial and momentum distributions, and the interaction potentials. In this section, we introduce the phenomenological potentials V_H which were described in Chap. 2 (Sect. 2.4.3), and will be applied either to the target and the projectile, or to the projectile only. In both cases, the initial distributions were chosen so that they were close to the quantum distributions. We have to recall that, in such a case, the electron energy on the target is not unique, but follows a distribution whose average value is –13.6 eV.

SC and SI cross sections are presented in Figs. 8.1 and 8.2 as a function of the projectile energy, and compared with experiment and theories described in Chap. 3.

We focus first on SC cross sections. Above 30 keV, the agreement between the present calculations and the experiment and previous theories is good. At lower energies, the best agreement is obtained when the two phenomenological potentials are included (full red curve). However, a disagreement of about 2 is again observed with the experiment. Unfortunately, no calculated result using previous models exist at these low energies. It would be necessary to see the behavior of the previous CTMC calculations (full black curve), since it seems that the SC cross sections probably decrease with decreasing E_P.

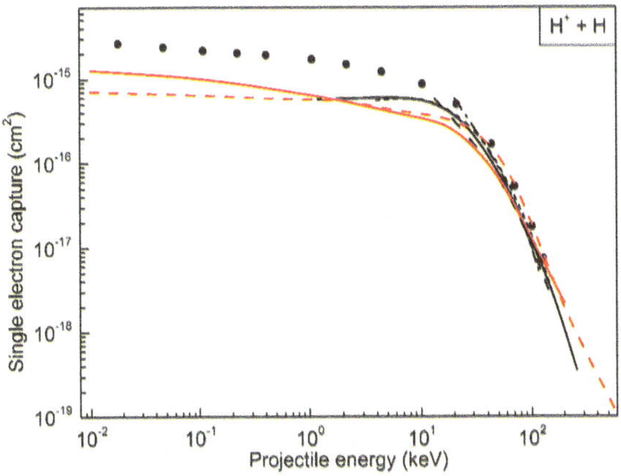

Fig. 8.1 SC cross sections following H^+ + H collisions, as a function of the projectile energy. Experiment, full circles; previous theories, blue curves (cf. Chap. 3); present CTMC calculations, red curves. Full red curve, the potentials V_H between the electron and the projectile and the electron and the target are included. Dashed red curve, only the potential V_H between the electron and the target is included

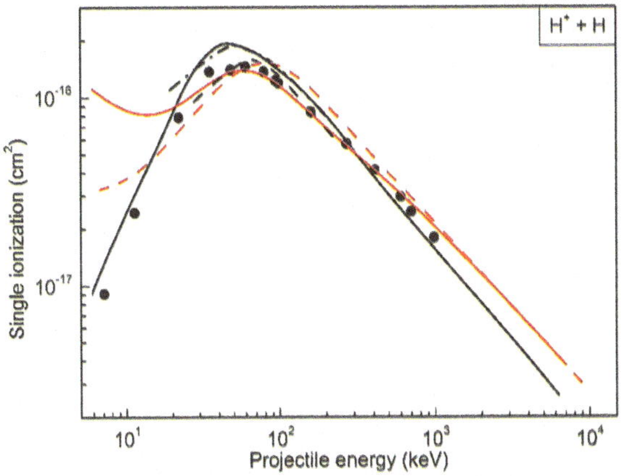

Fig. 8.2 SI cross sections following H^+ + H collisions, as a function of the projectile energy. Experiment, full circles; previous theories, blue curves (cf. Chap. 3); present CTMC calculations, red curves. Full red curve, the potentials V_H between the electron and the projectile and the electron and the target are included. Dashed red curve, only the potential V_H between the electron and the target is included

The inclusion of V_H in the determination of SI cross sections does not improve the cross sections. In fact, it appears that, at energies smaller than 20 keV, the present calculations diverge. This indicates that, at present, no classical model is able to predict what happens at low projectile energies.

8.2.2 Ne^{10+} + He Auger Effect Dependence onto the Projectile Energy

In Chap. 4, we discussed the possibility to describe classically the Auger effect which is a pure quantum process, by determining quantities $\tilde{\sigma}_P$ for a given inelastic process P whose dimensions are the same than that a cross section, at different times of the collision (cf. Chap. 4, Sect. 4.5.1). This technique was successfully applied to Ne^{10+} + He collisions at a projectile energy E_P of 150 keV, by introducing the electron correlation process at each step of the collision. It was possible to determine the average Auger width $\overline{\Gamma}_A$, and this value was found to be consistent with the experiment one. Thus, the success of this study allows us to go into further details.

The autoionizing and nonautoionizing double capture processes (*NADC* and *ADC*, respectively) were studied experimentally in a large range of projectile energies. It was clearly demonstrated that, in contrast with what happens at 150 keV, configurations of non equivalent electrons ($3lnl'$ with $n \geq 6$) are more and more populated when E_P decreases. Thus, an evolution of the value of $\overline{\Gamma}_A$ with E_P is expected, since it depends strongly onto the overlap of the wave functions associated with each electron. More precisely, individual values of $\Gamma^A_{nln'l'}$ decrease when n increases, since the overlap between the wave functions tend toward 0. Since the procedure to calculate $\overline{\Gamma}_A$ was easy to implement and leaded to a good agreement with the experiment, it is possible to extent the calculation at smaller values of E_P.

However, the so-called experimental width was deduced not only from the experiment, but also from calculations in order to evaluate autoionization cross sections $\sigma^A_{nln'l'^1L}$ for a given term $nln'l'^1L$. This quantity was calculated using an independent electron model, which is correct at 150 keV, but not at smaller energies (typically 1 keV or 10 keV). Consequently, the experimental width will be affected by the non inclusion of electron correlation, and will lead to a disagreement with the CTMC results. In this section, preliminary results are given, without going into the details.

8.2.2.1 Total Inelastic Cross Sections

Before analyzing the dependence in time of the Auger process for several projectile energies, it is necessary to validate the present method by calculating ionization and capture cross sections as a function of E_P and by comparing the results with available experimental and/or theoretical cross sections.

As an example, in Fig. 8.3, total cross sections for SI (short dashed red curves) and SC (full red curves) are presented, using a full body CTMC model (top of Fig. 8.3) and an independent electron CTMC model (bottom of the figure). The sum of these cross sections is depicted by the dashed red curves and compared with analytical form derived from experiment for single-electron loss (SC + SI) cross section [1] (dashed blue curve). For both models, the agreement is reasonable, since the behavior is reproduced, whereas the present CTMC models underestimate the cross sections at low energies by a factor of more than 3. However, it is recalled that the formula given

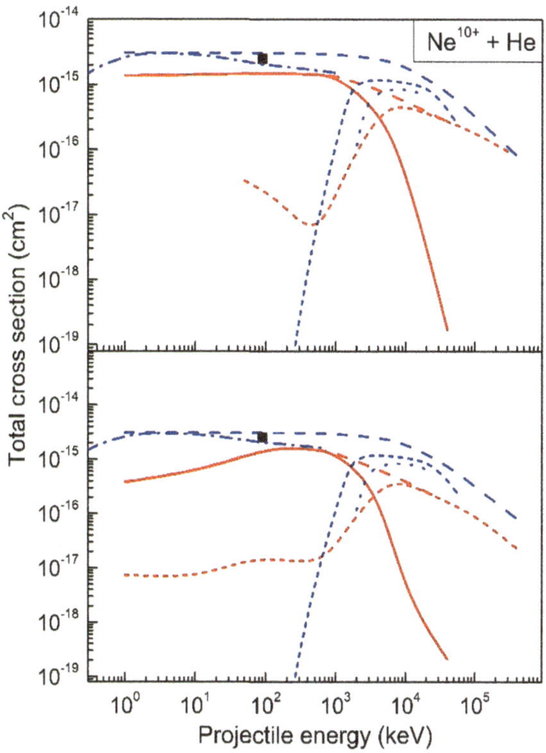

Fig. 8.3 Single electron-loss cross section following Ne^{10+} + He collisions as a function of the projectile energy E_P. C using quantum mechanics, full square; present CTMC calculations: full-body CTMC model, top of the figure, independent electron CTMC model, bottom of the figure; present SC and SI calculations, full and short dashed red curves, respectively; MLZ model, dotted dashed curves

by Olson was deduced for projectiles with charges smaller than 9. In addition, SI and SC cross sections are compared with previous results (short dashed and dotted blue curves for SI and dotted dashed curve for SC) [2, 3]. The agreement is reasonable, except for SI at energies smaller than 300 keV where SI cross sections seem to increase with decreasing E_P. Finally, it is noted that our results are consistent with the SC cross section calculated using quantum mechanics [4].

The NADC (short dashed red curves) and ADC (dashed red curves) cross sections are shown in Fig. 8.4 as a function of E_P. Again, results including electron correlation are presented on the top of Fig. 8.4, whereas results using independent electrons are presented on the bottom of the Fig. 8.4. The CTMC calculations are compared with previous experimental results (circles for ADC, open squares for NADC and full squares for the total DC cross section) in the range from 1 to 150 keV. The agreement is reasonable, even if NADC cross sections are overestimated, and is found to be better when electron correlation is taken into account. At large energies, DC calculations are in relatively good agreement with previous CTMC calculations (full blue curve), especially when electron correlation is not included.

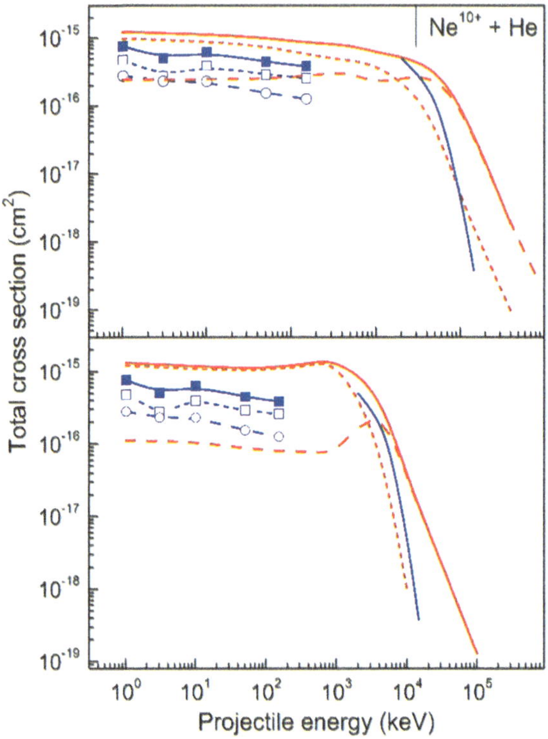

Fig. 8.4 Double capture cross sections following Ne^{10+} + He collisions as a function of the projectile energy E_P. Present model: *ADC, NADC* and total cross sections, dashed, short dashed and full red curves, respectively. Experiment: *ADC, NADC* and total cross sections, dashed blue curves and open circles, short dashed blue curves and open squares, full blue curves and full squares, respectively. Full blue curves, previous CTMC results. The present calculations are obtained by including electron correlation (top of the figure) or using an independent electron model (bottom of the figure)

8.2.2.2 Time Dependence of Auger Probabilities

Figures 8.5 and 8.6 represent the quantities $\tilde{\sigma}_{NADC}$ and $\tilde{\sigma}_{ADC}$ as a function of the integration time at $E_P = 1$, 10, 50 and 150 keV (empty symbols and dotted dashed, short dashed, dashed and full red curves, respectively). At 150 keV and 50 keV, the exponential decay for $\tilde{\sigma}_{NADC}$ and growth for $\tilde{\sigma}_{ADC}$ are clearly observed, whereas the decay and growth are less pronounced at smaller energies. It has to be recalled that the decay or growth is defined as soon as the two atoms are separated.

Note that the total cross section, at large integration time, changes strongly for *NADC* process, whereas that for *ADC* process is quite constant. This has to be compared with the available experimental and theoretical cross sections.

The calculated average Auger width is deduced from a fitting procedure described previously, and compared with the experimental Auger width (Fig. 8.7). The agreement between both values is remarkable, even if the electron correlation could not be

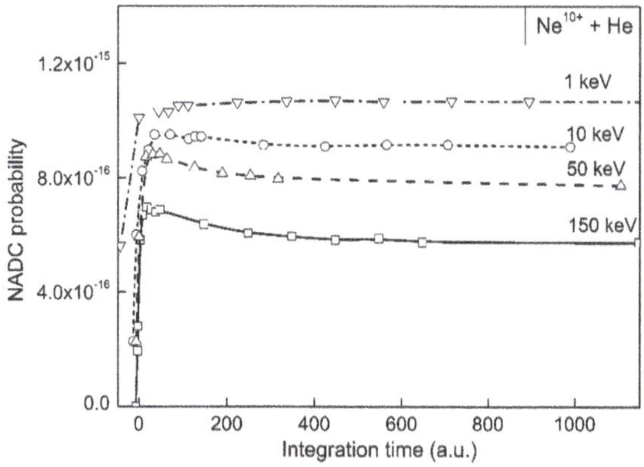

Fig. 8.5 Results of the CTMC calculations for the quantity $\tilde{\sigma}_{NADC}$ following Ne^{10+} + He collisions at projectile energies $E_P = 1$, 10, 50 and 150 keV, as a function of the integration time

Fig. 8.6 Results of the CTMC calculations for the quantity $\tilde{\sigma}_{ADC}$ following Ne^{10+} + He collisions at projectile energies $E_P = 1$, 10, 50 and 150 keV, as a function of the integration time

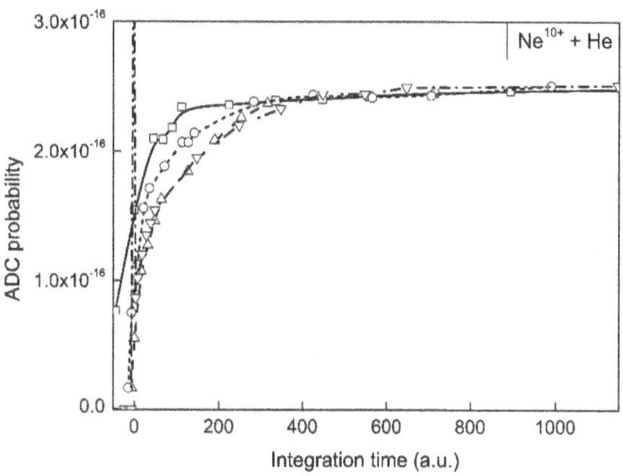

Fig. 8.7 Calculated (full red curve) and experimental (dashed black curves) average Auger widths following Ne^{10+} + He collisions as a function of the projectile energies E_P

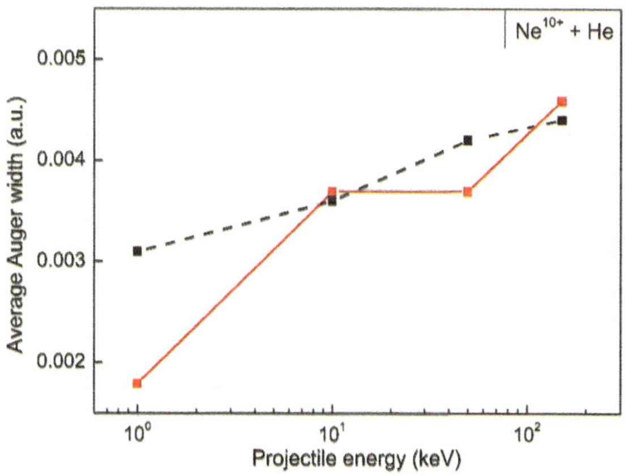

taken into account in the determination of L experimental distributions. This encourages us to deepen several points. For example, impact parameters can be calculated and compared with other theories, especially quantum theories. Cross sections for a given process including or not including electron correlation can also be compared in order to enlighten the role of electron correlation. Finally, electron energy distributions can also be studied as a function of E_P and compared with experiment. Moreover, for a fixed value of E_P, the evolution in time of the electron energy distributions can be followed.

8.3 Classical Description of the H_2 Molecule Including Electron Correlation

To our knowledge, no theoretical classical calculations were performed for ion-H_2 collisions using Coulombic potentials for the two protons and both electrons. The problem to solve is much more difficult to solve than that in the case of *He* atom, due to the poor stability of the molecule, especially when the collision occurs at very low impact energies. Moreover, several combinations in the initial conditions are possible to build the molecule. Thus one would have to check these possibilities in order to improve the final results. Among all these configurations, Fig. 8.8 presents some of them. The protons and the electrons are represented by green and red spheres, respectively, and the possible trajectories are depicted by blue curves. The most important points are that, whatever the chosen initial configurations:

- The electrons are located symmetrically with respect to the proton axis.
- Their initial velocities are equal in intensity but the associated vectors are opposite.
- During their movement within the molecule in the absence of any projectile, no singularity should appear.

As mentioned in Chap. 6, the only configuration that was used to treat the collision was that depicted in Fig. 8.8 (top left). No other configuration was checked. It follows that a systematic study for the H_2 molecule would be useful, especially to follow what happens at very small projectile energies. Indeed, at present, at small energies, the SI cross sections were found to increase, which is at odd with the experimental observations.

8.4 Double Capture Following $He^{2+} + H_2$ Capture

Among all the collision systems, the $He^{2+} + H_2$ system is of great interest since this is the simplest one in which double events, such as DI, TI or DC, may occur. This system was explored in details experimentally and theoretically, in a wide range of collision energies [5–9]. In the specific case of DC process, the experimental and

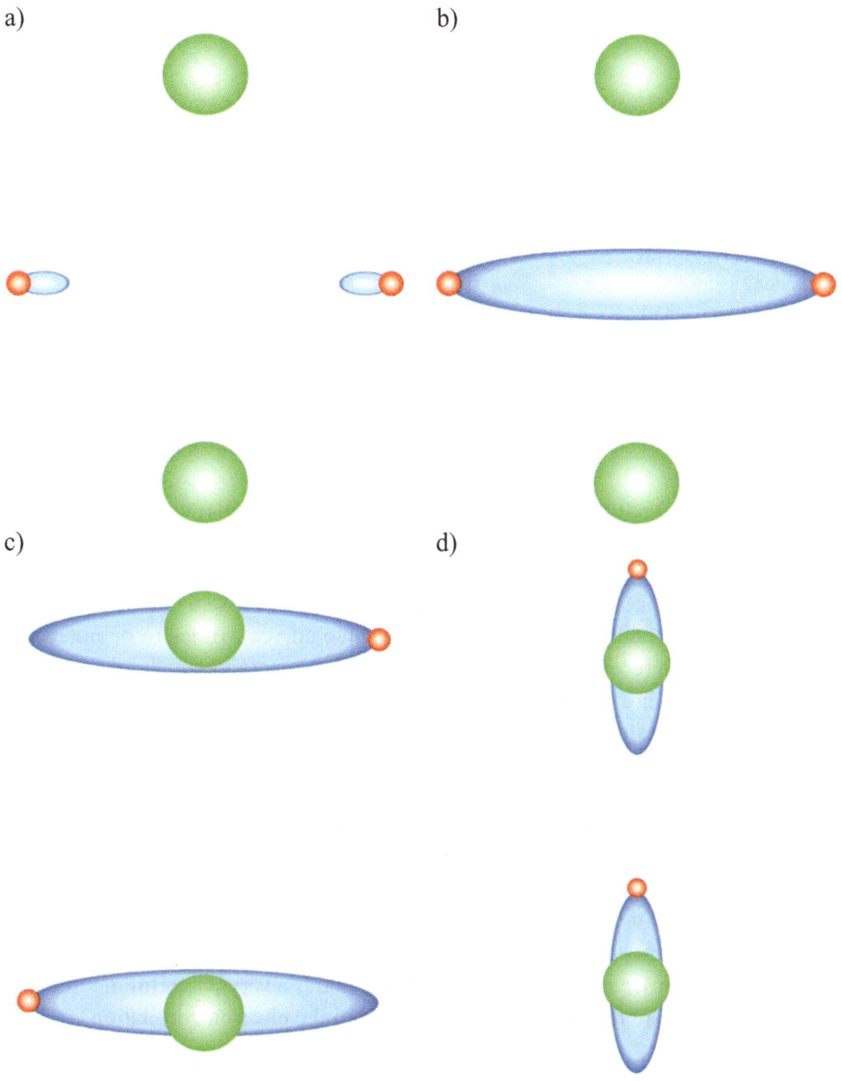

Fig. 8.8 Schematic view of some possible initial configuration of the H_2 molecule. Protons, green spheres; electrons, red spheres; electron trajectories, blue curves

theoretical results are coherent, but some discrepancies occur, especially at impact energies smaller than 50 keV. Consequently, the CTMC method would be useful to check the model at any collision energies. Preliminary calculations were made using independent and dependent electrons.[1]

[1] Not yet published.

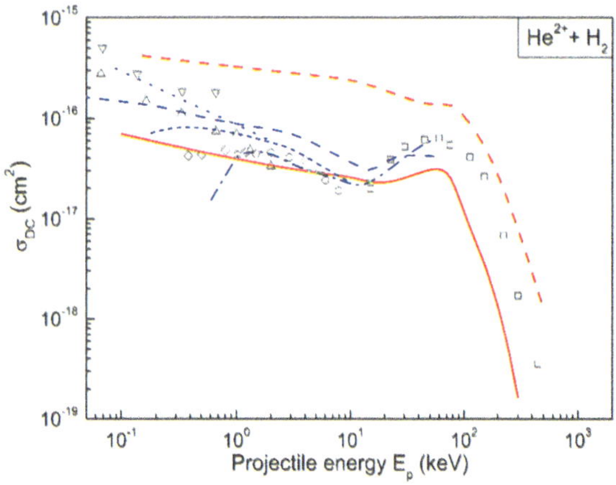

Fig. 8.9 DC capture cross sections following He^{2+} + H_2 collisions as a function of the projectile energy. Experiment, empty objects (see [9] and ref. therein); previous calculations, blue curves; CTMC calculations using independent electrons, dashed red curve; CTMC calculations using correlated electrons, full red curve

The DC cross sections are presented in Fig. 8.9 as a function of the projectile energy. The experimental cross sections (empty triangles, diamonds, circles and squares) present a maximum at about 75 keV and a minimum at about 10 keV. At energies smaller than 2 keV, the experimental data are not consistent with each other. Some theoretical results are also represented (blue curves). It has to be noted that only cross sections below 100 keV were calculated. At energies smaller than 10 keV, strong divergences are seen between the theories themselves and between theoretical and experimental results.

The CTMC results are represented by dashed and full red curves, when independent and correlated electrons are used, respectively. The independent electron model overestimates the experiment by a factor of approximately 2, and the maximum and the minimum are not observed. In contrast, the dependent electron model underestimates the experiment at high energies, but nicely reproduce some experimental results below 10 keV. In addition, the maximum is clearly seen, and located at the same energy than that observed experimentally.

These results prove again that the CTMC method is a very powerful method, and that many studies are required to take into account the electron correlation.

How to introduce electron correlation in the classical treatment of ***H_2O ?***

It seems, at first glance, impossible to introduce electron correlation for molecules such as H_2O, since this molecule has less axes of symmetry than atoms or molecules such as He or H_2. One would expect that the introduction of electron correlation would lead to strong instabilities, thus to spontaneous ionization, due to the presence of a projectile.

Nevertheless, in first approximation, it would be possible to introduce the electron correlation step by step. The H_2O molecule consists of three nuclei and ten electrons, two of which are completely bound to *O* nucleus. These two electrons can be considered as frozen. Consequently, eight electrons have to be taken into account. Figure 8.10 presents a possible configuration that, in principle, can lead to stable molecules. The nuclei interact with each other and all the electrons using Coulombic

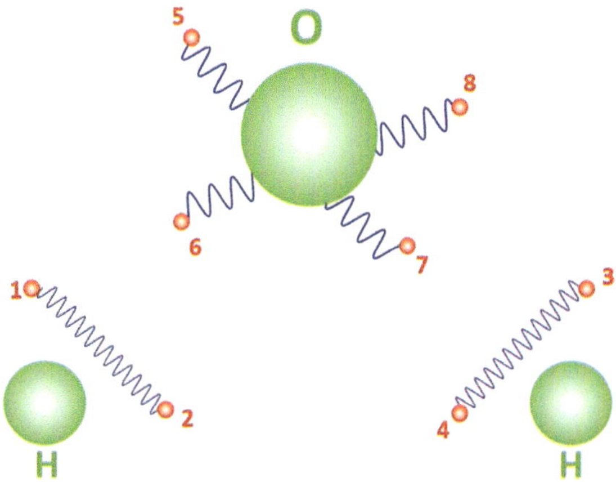

Fig. 8.10 Schematic view of a H_2O molecule. The nuclei and the electrons are depicted by green and red circles, respectively. The blue links between electrons represent the electron–electron interactions

potentials, as usual. The electrons are linked by pairs (1–2, 3–4, 5–6, 7–8) in order to avoid chaotic trajectories.

So far, no theoretical study using CTMC method has been conducted for such complicated molecule. This study is important since it is a complete new way to interpret the results, especially cross sections and fragmentation.

References

1. Olson, R.E.: Electron-capture and impact-ionization cross sections for multiply charged ions colliding with helium. Phys. Rev. A **18**, 2464 (1978)
2. Lyons, D., Cumbee, R.S., Stancil, P.C.: Charge exchange of highly charged Ne and Mg ions with H and He. The Astrophy. J. Supll. Series **232**, 1 (2017)
3. Wetmore, A.E., Olson, R.E.: Electron loss from helium atoms by collisions with fully stripped ions. Phys. Rev. A **38**, 5563 (1988)
4. Leung, A.C.K., Kirchner, T.: Independent-electron analysis of the x-ray spectra from single-electron capture in Ne^{10+} collisions with He, Ne and Ar. Phys. Rev. A **92** 032712 (2015)
5. Shah, M.B., Gilbody, H.B.: Single and double ionization of helium by H^+, He^{2+} and Li^{3+} ions. J. Phys. B **18**, 899 (1985)
6. Okuno, K., Soejima, K., Kaneko, Y.: Single- and double-electron capture in $^3He^{2+}$+ H_2 collisions. J. Phys. B **25**, L105 (1992)
7. Errea, L.F., Macias, A., Méndez, L., Pons, B., Riera, A.: Molecular treatment of single (dissociative and non dissociative) and double electron capture in $He^{2+} + H_2$ collisions. J. Phys. B **36**, L135 (2003)
8. Figueira da Silva, S., Winter, H.P., Aumayr, F.: Single- and double-electron capture cross sections for slow He^{2+} impact on O_2, H_2 and D_2. Phys. Rev. A **75**, 042706 (2007)
9. Sisourat, N., Pilskog, I., Dubois, A.: Non perturbative treatment of mulitelectron processes in ion-molecule scattering: Application to He^{2+}+ H_2 collisions. Phys. Rev. A **84**, 052722 (2011)

Appendix A
Extensions of the Classical Over-Barrier Model

A.1 Charge Exchange Model for Atoms

A.1.1 Motivation

A charge exchange model has been developed [1] in order to treat the particular case of collisions of singly or multiply charged ions with atoms having a weakly bound active electron. The present model offers the possibility of calculating charge exchange probabilities and total cross sections by using an analytical formula that is expected to be valid at low and intermediate projectile velocities.

A.1.2 Description of the Model

Suppose an active electron that is molecularized, in the field of an projectile of charge q and an atomic target. The potential energy is

$$V(r, R) = -\frac{1}{r} - \frac{q}{\| \vec{r} - \vec{R} \|} \tag{A.1}$$

where r is the distance between the electron and the projectile nucleus, and R is the internuclear distance (Fig. A.1). Initially, the electron has a binding energy E_{n}. At a distance R, the electron energy $E(R)$ is modified by the presence of the projectile, so that

$$E(R) = -E_n - \frac{q\xi}{R} \tag{A.2}$$

© The Editor(s) (if applicable) and The Author(s), under exclusive license to Springer Nature Switzerland AG 2021
F. Frémont, *Classical Treatment of Collisions Between Ions and Atoms or Molecules*, Springer Series on Atomic, Optical, and Plasma Physics 118,
https://doi.org/10.1007/978-3-030-89428-3

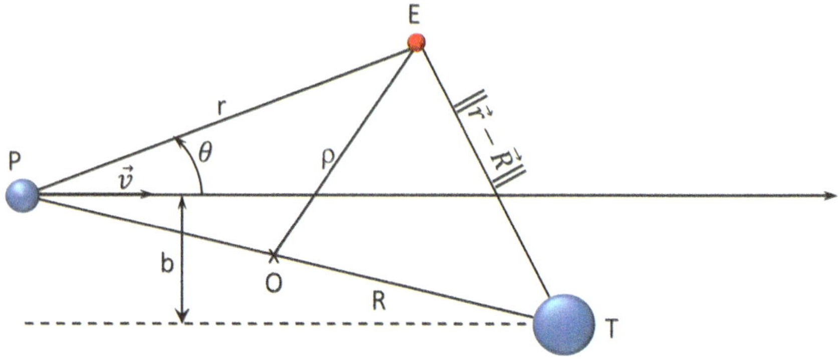

Fig. A.1 Schematic view of the collision between a projectile P and a target T. The electron E is shared between the two nuclei. O is the middle of PT. The electron is defined by ρ and θ

where ξ is a parameter in the range from 0 to 1, that characterizes the deepness of the barrier at its maximum. In the following, the value of 1 will be chosen for ξ.

As in the case of the COB model (Chap. 2), electron capture occurs when $E(R) = V(r, R)$. We call ρ the electron cylindrical coordinate. The maximum value ρ_m at which capture occurs is given by

$$E_n + \frac{q\xi}{R} = \left[\rho_m^2 + \frac{R^2}{\left(\sqrt{q}+1\right)^2}\right]^{-\frac{1}{2}} + q\left[\rho_m^2 + \frac{qR^2}{\left(\sqrt{q}+1\right)^2}\right]^{-\frac{1}{2}} \tag{A.3}$$

which cannot be analytically solved, except for $q = 1$.

Expanding expression (A.3) in powers of ρ_m, we obtain

$$\rho_m^2 \approx \frac{2\sqrt{q}}{\left(\sqrt{q}+1\right)^4}\left[\left(\sqrt{q}+1\right)^2 - E_n R - q\xi\right] \tag{A.4}$$

Thus, the corresponding value θ_m for θ, given by

$$\cos\theta_m \approx \frac{R}{2\left(\rho_m^2 + \frac{R^2}{4}\right)^{1/2}} \tag{A.5}$$

can be determined in analytic form

$$1 - \cos\theta_m \approx \frac{\sqrt{q}}{\left(\sqrt{q}+1\right)^2}\left[\left(\sqrt{q}+1\right)^2 - E_n R - q\xi\right] \tag{A.6}$$

The probability W on finding an electron bound on the target after the collision decreases according to

$$dW = -N_\Omega \frac{dt}{T} W \quad \text{(A.7)}$$

where T is the electron motion period and N_Ω is the ratio between the number of trajectories N_C giving rise to capture and the total number N of trajectories

$$N_\Omega = \frac{1}{4\pi} \int_0^{2\pi} d\varphi \int_0^{\theta_m} \sin\theta d\theta \quad \text{(A.8)}$$

Thus, the electron capture probability is

$$P(b, v) = 1 - W(\infty) = 1 - \exp\left[\frac{1}{T} \int_{-\infty}^{+\infty} N_\Omega dt\right] \quad \text{(A.9)}$$

Since $R = \sqrt{b^2 + v^2 t^2}$ if we assume straight line trajectories for the projectile, $P(b, v)$ is easily calculated

$$P(b, v) = 1 - \exp\left[\frac{1}{8vT}\left\{\left[\left(\sqrt{q}+1\right)^2 - q\xi\right] z_m - E_n b^2 \ln\frac{z_m + \left(\sqrt{q}+1\right)^2 - q\xi}{b}\right\}\right] \quad \text{(A.10)}$$

where

$$z_m = \left[\frac{\left(\sqrt{q}+1\right)^2 - q\xi}{E_n^2} - b^2\right]^{1/2} \quad \text{(A.11)}$$

Finally, total cross section is determined by using the relation

$$\sigma_C = 2\pi \int_0^{b_{max}} bP(b, v) db \quad \text{(A.12)}$$

A.1.3 Applications

A.1.3.1 O^{8+} + H Collisions

The present model is typically not suitable for O^{8+} + H collisions, since the 1s electron of H is tightly bound. However, despite the restriction, it has been applied to

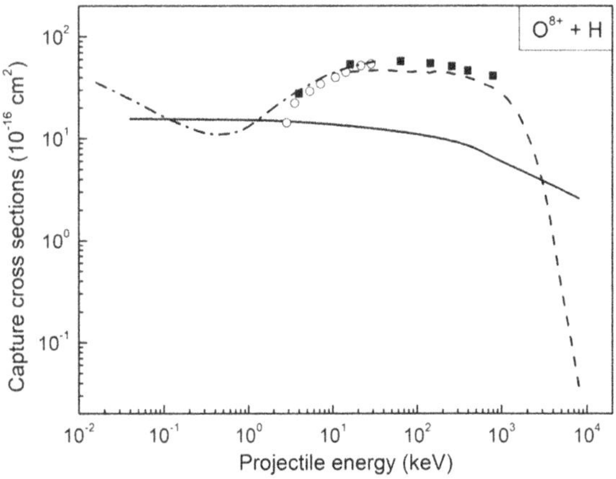

Fig. A.2 Total capture cross sections following $O^{8+} + H$ collisions as a function of projectile energy. Experiment: full squares [2], empty circles [3]; calculations: dashed curve [4], dotted-dashed curve [5], full curve, Ostrovsky's calculations [1]

such collision system. Figure A.2 shows experimental cross sections (squares [2] and circles [3]), compared to various calculations (dashed [4] and dotted-dashed curves [5]), as well as calculations originating from Ostrovsky's theory (full curve).

Cross sections derivated from Ostrovsky's model are of the same order of magnitude as those found theoretically at small projectile energies; but noticeably deviate from experiment and theory at intermediate and high energies.

A.1.3.2 $A^{q+} + Cs(6s)$ Collisions

$A^{q+} + Cs(6s)$ collisions are typical collisions for which Ostrovsky's model is expected to give good results. Figure A.3. compares, for N^{5+} and Ar^{3+} projectiles, capture prob-

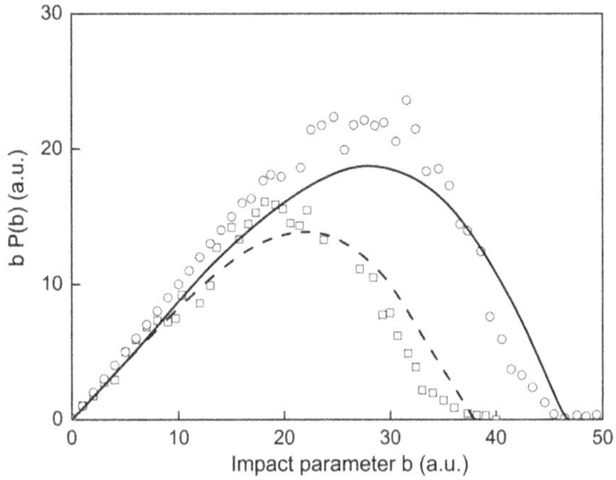

Fig. A.3 Calculated probability of electron capture as a function of the impact parameter b in $N^+ + Cs(6\,s)$ collisions at $v = 0.894$ a.u. (open squares, CTMC calculations [6]; dashed curve, present model [1]) and in $Ar^+ + Cs(6s)$ collisions at $v = 0.894$ a.u. (open circles, CTMC calculations [6]; full curve, present model [1])

abilities as a function of impact parameter. The present calculations are compared with CTMC calculations [6]. Keeping in mind that the present model is extremely simple, the agreement is remarkable, in shape and in absolute scale. The parameter for ξ was chosen to be equal to 1.

A.2 Angular Momentum Distributions of Electron Capture

A.2.1 Goal of the Model

During the collision, the projectile of velocity v captures one electron at a certain impact parameter b. Consequently, in average, the captured electron gains angular momentum $\updownarrow$ of the order of bv. As an example, at $v = 0.5$ a.u. and for $b = 8$ a.u., one obtain $\updownarrow = 4$. This induces an electron captured onto n = 5 state of the projectile. The transfer is an important feature of a collision, especially when more than one electron are involved, so that a classical development of $\updownarrow$ distributions is necessary to calculate.

A.2.2 Determination of Angular Momentum Distributions

A preliminary model was created [7] to determine the average angular momentum of a captured electron in slow collisions as a function of projectile velocity v. Since COB model is static by nature, v was introduced in the expression of capture probability.

The model is identical to that formulated by Niehaus (cf. Chap. 2), except for an additional term in the expression of the potential $V(x)$, that takes into account angular momentum

$$V(x) = -\frac{q}{x} - \frac{1}{\| \overrightarrow{R} - \overrightarrow{x} \|} + \frac{L^2}{2x^2} \tag{A.13}$$

where L is the classical angular momentum.

With the condition of energy resonance

$$I_t - \frac{q}{R} = \frac{-q^2}{2n^2} - \frac{1}{R} \tag{A.14}$$

where $I_t = 0.5$ a.u. in the case of an electron onto a hydrogen target.

The expressions for capture distance $R_C(L)$ and principal quantum number $n(L)$ could be derived

$$R_C(L) \leq \left(2\sqrt{q}+1\right)\left[1+\left(1-\frac{\left(1+\sqrt{q}\right)^2 L^2}{\left(2\sqrt{q}+1\right)^2 q}\right)^{1/2}\right] \tag{A.15}$$

$$n(L) = q\left(\frac{2\sqrt{q}+1}{q+2\sqrt{q}}\right)^{1/2}\left[1+\frac{(q-1)\left(1+\sqrt{q}\right)^2 L^2}{4\left(2\sqrt{q}+1\right)^3 q}\right]^{-1/2} \tag{A.16}$$

L is connected to the quantum value of angular momentum by $L^2 = \updownarrow(\updownarrow + 1)$. Consequently, for a given collision, L reaches a maximum value since $\updownarrow \leq n - 1$.

In this model, capture cross section for a given L writes

$$\sigma_L = A2\pi \int_0^{R_C(L)} H(L+1-bv)H(bv-L)bdb \tag{A.17}$$

where H are Heaviside distributions.

The average angular momentum is given by

$$\langle L\rangle = \left(\sum_{L=0}^{n-1}(L+1)\sigma_L\right) / \sum_{L=0}^{n-1}\sigma_L \tag{A.18}$$

The quantity $\langle L\rangle$ was represented as a function of v for ion-hydrogen collisions and extended to $N^{5+} + H_2$ collisions. As an example, $\langle L\rangle$ distribution is shown in Fig. A.4 for $O^{6+} + H$ collisions and Fig. A.5 for $N^{5+} + H_2$ collisions (full curve) and compared to experimental distribution (open circles) [8]. In the case of atomic hydrogen target, theoretical results match very well experiment. In contrast, for dihydrogen target, theoretical results under estimate experiment. No reason was given by the authors, however we suggest that this is partly due to the fact that, at low energy, both target

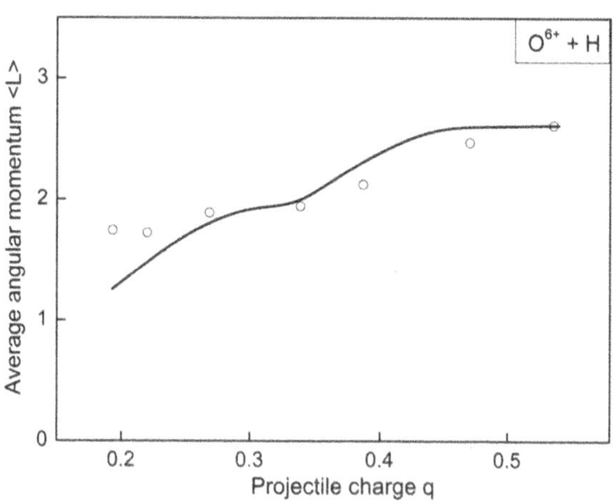

Fig. A.4 Average angular distribution obtained experimentally (open circles) [8] and theoretically (full curve) [7] as a function of projectile velocity for O^{6+} + H collisions

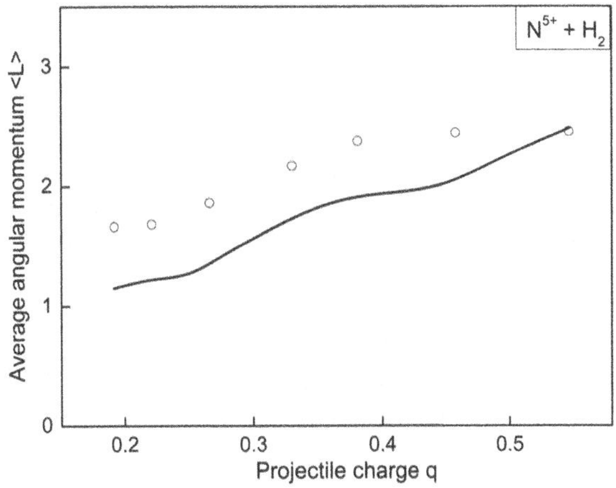

Fig. A.5 Average angular distribution obtained experimentally (open circles) [8] and theoretically (full curve) [7] as a function of projectile velocity for N^{5+} + H_2 collisions

electrons cannot be considered as independent. Electron correlation has to be taken into account, and would probably give correction to the L distributions.

Improvements of the model were established one year later [9]. Instead of taking Heaviside distributions, the authors used Gaussian distributions in energy, leading to Gaussian distributions in L, due to the fact that, even or well defined values of b and v, L is defined with an uncertainty ΔL.

The capture cross section for a given L writes

$$\sigma_L = \frac{2L+1}{2L+2}\frac{2\sqrt{\pi}}{\Delta L}\int_0^{R_C(L)}\left[exp - \left(\frac{L+\frac{1}{2}-bv}{\Delta L}\right)^2 + exp - \left(\frac{L+\frac{1}{2}+bv}{\Delta L}\right)^2\right] bdb \tag{A.19}$$

Figure A.6 shows capture cross sections for O^{6+} + H collisions as a function of v, for $L = 0$, 1, 2 and 3. Capture derived from relations (A.17) (dashed curves) and (A.18) (full curves) are compared with experimental partial cross sections (empty circles) [8]. In the case of largest values of L, Gaussian distributions improve the agreement with experiment. For s and p states, no clear improvement of the model is observed, except perhaps for $L = 1$.

Calculated total cross sections are also represented (dashed curve and full curve for Heaviside and Gaussian L distributions, respectively) and compared with previous experiment [8] (Fig. A.7). Discrepancies between calculations and experiment are clearly visible, since experimental cross sections are constant, while calculated cross sections strongly increase with decreasing v. The oscillations observed in the case of Heaviside distributions are due to the discontinuity of such distributions.

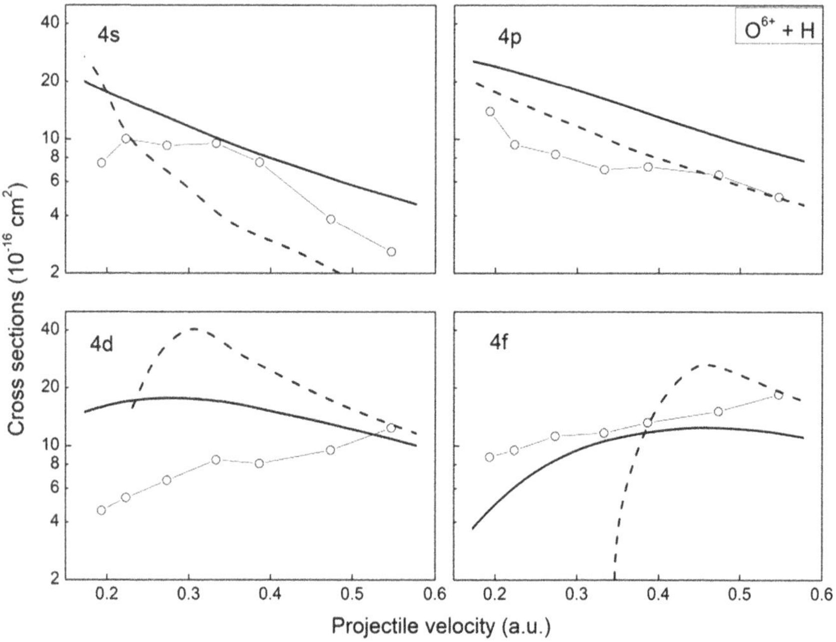

Fig. A.6 Angular momentum distributions following single capture in O^{6+} + H collisions as a function of the projectile velocity v. Experiment, empty circles [8]; calculations with Heaviside distributions, dashed curves; calculations using Gaussian distributions, full curves [9]

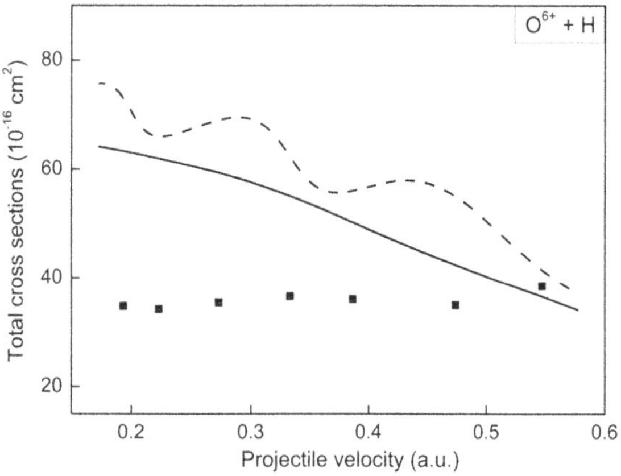

Fig. A.7 Total capture cross sections following O^{6+} + H collisions as a function of the projectile velocity v. Experiment, empty circles [8]; calculations with Heaviside distributions, dashed curves; calculations using Gaussian distributions, full curves [9]

References

1. Ostrovsky, V.N: Rydberg atom-ion collisions: classical overbarrier model for charge exchange. J. Phys. B **28**, 3901 (1985)

2. Bendahman, M., Bliman, S., Dousson, S., Hitz, D., Gayet, R., Hanssen, J., Harel, C., Salin, A.: Electron capture from atomic hydrogen by multiply charged ions in low energy collisions. J. Phys. **46**, 561 (1985)
3. Meyer, F.W., Howald, A.M., Havener, C.C., Phaneuf, R.A.: Low-energy total-electron-capture cross sections for fully stripped and H-like projectile incident on H and H_2. Phys. Rev. A **32**, 3310 (1985)
4. Lee, T.-G., Hesse, M., Le, A.-T., Lin, C.D.: Charge transfer in slow collisions of O^{8+} and with H(1s) below 2 keV/amu. Phys. Rev. A **70**, 012,702 (2004)
5. Pandley, M.K., Lin, Y.-C., Ho, Y.K.: Cross sections of charge exchange and ionization in $O^{8+} + H$ collisions in Debye plasmas. Phys. Plasmas **19**, 062,104 (2012)
6. Pascale, J., Olson, R.E., Reinhold, C.O.: Phys. Rev. A **42**, 5305 (1990)
7. Burgdörfer, J., Morgenstern, R., Niehaus, A.: Angular momentum distribution in the classical over-barrier model for electron capture into highly charged slow projectiles. J. Phys. B **19**, L507 (1986)
8. Dijkkamp, D., Ciric, D., Vlieg, E., de Beer, A., de Heer, F.J.: Subshell-selective electron capture in collisions of C^{4+}, N^{5+}, C^{6+} with H, H_2 and He. J. Phys. B **18**, 4763 (1985)
9. Burgdörfer, J., Morgenstern, R., Niehaus, A.: Angular momentum of captured electrons: the classical over-barrier model and its limitations. Nucl. Instr. Meth. Phys. Res. **B23**, 120 (1987)

Appendix B
Classical Radial and Momentum Distributions

B.1 Continuous Distributions with Variable Energies

B.1.1 Wigner Distributions

In 1932, E. Wigner proposed a general form for the spatial and momentum distribution, in such manner that they are similar to quantum distributions [1] (For complete details, see [2]). Indeed, to treat $^+ + H$ collisions, the microcanonical distribution leads to a lack of ionization, excitation and charge transfer at large impact parameters b. Thus, it is necessary to make the radial distribution wider. For this reason, a distribution $f_W(\vec{r}, \vec{p})$ was introduced to reproduce quantum distributions. Its expression is

$$f_W(\vec{r}, \vec{p}) = \frac{1}{\pi^4} \int \exp(2i\vec{p}.\vec{y})\exp(-|\vec{r} - \vec{y}| - |\vec{r} + \vec{y}|)d^3\vec{y} \quad \text{(B.1)}$$

with $\vec{r}\,(0, 0, r)$, $\vec{p}\,(0, p\sin\alpha, p\cos\alpha)$ and $\vec{y}\,(y\sin\theta\cos\varphi, y\sin\theta\sin\varphi, y\cos\theta)$.

The energy distribution is given by

$$\rho_W(E) = \int f_W(\vec{r_e}, \vec{p_e})\delta\left(E - \frac{p_e^2}{2} - \frac{1}{r_e}\right)d^3\vec{r_e}\,d^3\vec{p_e} \quad \text{(B.2)}$$

and its form is shown in Fig. B.1.

This distribution cannot be used in the present form for two reasons:

- $\rho_W(E)$ is negative in the energy range from 0.86 a.u. to 1.29 a.u. Consequently, $\rho_W(E)$ cannot be interpreted as a density.
- Some values of r and p give rise to positive energies. For these particular values, the electron is no more bound on the target.

© The Editor(s) (if applicable) and The Author(s), under exclusive license to Springer Nature Switzerland AG 2021

F. Frémont, *Classical Treatment of Collisions Between Ions and Atoms or Molecules*, Springer Series on Atomic, Optical, and Plasma Physics 118,
https://doi.org/10.1007/978-3-030-89428-3

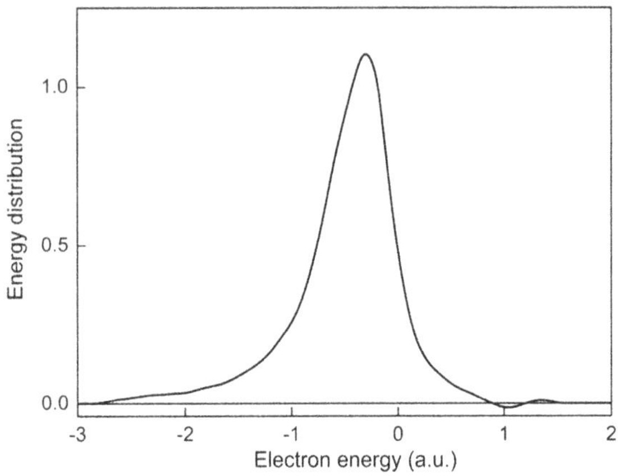

Fig. B.1 Energy distribution of the electron bound on the fundamental state of H using Wigner function

The two non-physical conditions $\rho_W(E) < 0$ and $E > 0$ can be simultaneously suppressed if the Wigner distribution is cut off for $E > E_{max}$, with $E_{max} \leq 0$. The new distribution writes

$$F_W\left(\overrightarrow{r_e}, \overrightarrow{p_e}\right) = \begin{cases} \frac{1}{N} f_W\left(\overrightarrow{r_e}, \overrightarrow{p_e}\right) for\, E_{min} \leq E \leq E_{max} \\ 0 elsewhere \end{cases} \tag{B.3}$$

The cut-off Wigner distributions yield only approximation of radial and momentum distributions, as shown in Figs. B.2 and B.3. However, the radial Wigner distribution (full curve in Fig. B.2) is now close to the quantum one. The maximum of the Wigner distribution is shifted to larger distance, and the width is smaller than the quantum one (dashed curve). The momentum Wigner distribution (full curve in Fig. B.3) is in rather good agreement with the quantum one (dashed curve).

As it will be seen in Chap. 3, the present distribution improves ionization and charge exchange cross sections for the $^+ + H$ collision system in the intermediate and high projectile energy range.

B.1.2 Cohen and Sattin Distributions

In contrast with Wigner distribution, the distributions presented here depends only on the binding energy, yielding the exact distribution as the quantum one and providing a good approximation to the momentum distribution.

The problem to solve is to obtain the better form for energy distribution $f(E)$ that satisfies the following two integral equations

$$4\sqrt{2}\pi \int_{-1/r}^{0} f(E)\left(E + \frac{1}{r}\right)^{1/2} dE = \frac{e^{-2r}}{\pi} \tag{B.4}$$

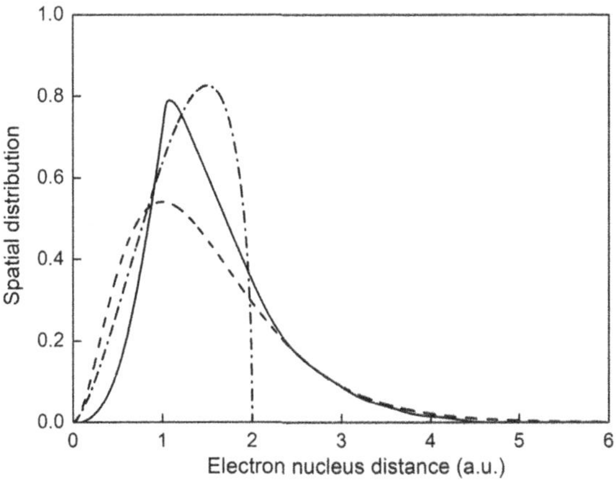

Fig. B.2 Classical spatial distributions of the 1s electron of H. Quantum distribution, dashed curve; microcanonical distribution, dotted-dashed curve; Wigner distribution, full curve

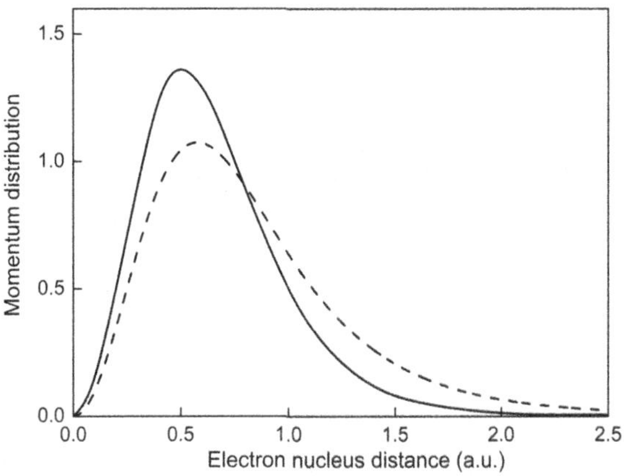

Fig. B.3 Classical momentum distributions of the 1s electron of H. Quantum distribution, dashed curve; Wigner distribution, full curve. The microcanonical distribution is identical to the quantum distribution

$$4\pi \int_{-1/r}^{0} f(E)\left(\frac{p^2}{2} - E\right)^{-4} dE = \frac{8}{\pi^2\left(1 + p^2\right)^2} \tag{B.5}$$

Equation B.4 can be solved analytically, given rise to the following expression for $f(E)$

$$f(E) = \frac{e^{1/E}}{2^{3/2}\pi^{5/2}(-E)^{3/2}}\left[\frac{1}{4}W_{-\frac{1}{2},-\frac{1}{2}}\left(\frac{-2}{E}\right) - \left(1 + \frac{2}{E}\right)W_{\frac{1}{2},-\frac{1}{2}}\left(\frac{-2}{E}\right)\right] \tag{B.6}$$

where W are Whittaker functions.

The energy density is given by

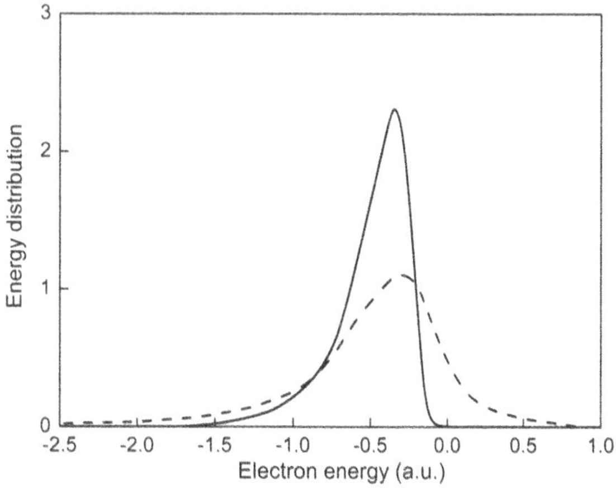

Fig. B.4 Energy distribution of hydrogen electron in its fundamental state using Wigner (dashed curve) and Cohen (full curve) radial distributions

$$\rho_E(E) = \sqrt{2}\pi^3 \frac{f(E)}{(-E)^{5/2}} \tag{B.7}$$

and is presented in Fig. B.4. The distribution (full curve) is narrower than the Wigner's distribution, and has the advantage to fall off very rapidly at $E > -0.25$ a.u.

Using expression (B.7), the momentum distribution

$$\rho_p(p) = \frac{256}{\pi^{3/2} p^{10}} \int_0^\infty x^{7/2} \left(\frac{4}{p^2} + x \right)^{-5} W_{\frac{1}{2},-\frac{1}{2}}(x) dx \tag{B.8}$$

is found to be very close to the quantum distribution (Fig. B.5). A shift in the maximum of only 0.02 a.u. is observed.

The latter distribution was applied to $^+ + H$ collisions (cf. Chap. 3) and give remarkable results for ionization and electron capture at projectile energies larger than 20 keV.

An improvement of the present model was provided by F. Satin, suggesting an alternative recipe based on the construction of stationary classical spatial distributions for electrons in arbitrary levels, on the target as well as on the projectile [3]. Consequently, the novelty is that the new expression for $f(E)$ is used to identify initial and final states. The energy density fulfils

$$\rho_E(E) = \sqrt{2}\pi^3 Z^3 \frac{f(E)}{(-E)^{5/2}} \tag{B.9}$$

where Z is the projectile or target charge.

The exact forms of $\rho_E(E)$ involve an increasing number of hypergeometric functions as the principal quantum number n increases. Consequently, more manageable expressions were found for any value of n

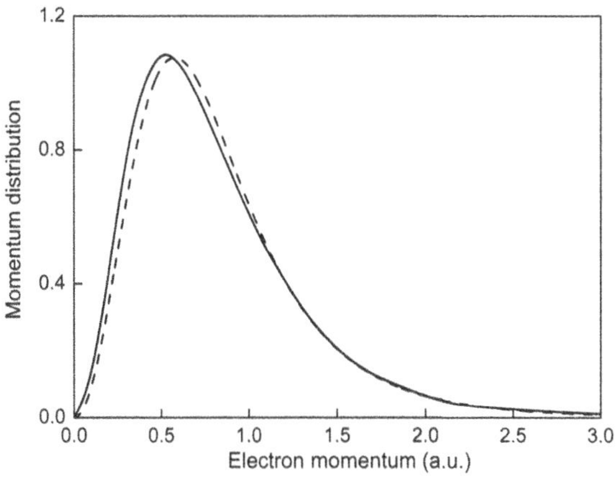

Fig. B.5 Momentum distribution of hydrogen electron in its fundamental state using quantum mechanics (dashed curve) and Cohen (full curve) distributions

$$\rho_E(E) = \frac{1}{\Gamma(6n)\beta^{6n}E^{6n+1}}\exp\left(-\frac{1}{\beta E}\right) \tag{B.10}$$

with

$$\beta = \frac{n+1}{5Z^2} \tag{B.11}$$

Using relation (B.10), the radial distribution can be analytically written

$$\rho(r) = \frac{4}{\Gamma(6n)\pi^2(\beta Z)^5\sqrt{x}}\left[\begin{array}{c}\Gamma\left(6n-\frac{5}{2}\right){}_1F_1\left(-\frac{1}{2},\frac{7}{2}-6n,-x\right)+\\ x^{5/2}\sqrt{\frac{\pi}{4}}\frac{\Gamma\left(\frac{5}{2}-6n\right)}{\Gamma(4-6n)}{}_1F_1\left(6n-3,6n-\frac{3}{2},-x\right)\end{array}\right] \tag{B.12}$$

This method was applied successfully to electron capture into different states for $He^{2+} + H$ and $Be^{4+} + H$ collisions.

B.2 Quasi-Continuous Radial Distributions with Discrete Energies

In contrast with the above models, the present one is based on series of microcanonical distributions, in such manner that the binding energy E is varied to allow the electronic radial distribution to extend beyond 2 a.u. [4]. The radial function $\rho_E(r_e)$ was obtained by requiring that a discrete set of radial distributions $\rho_E(r_e, E_i)$ reproduces the quantum one

$$\rho_E(r_e) = \sum_i a_i \rho_E(r_e, E_i) \tag{B.13}$$

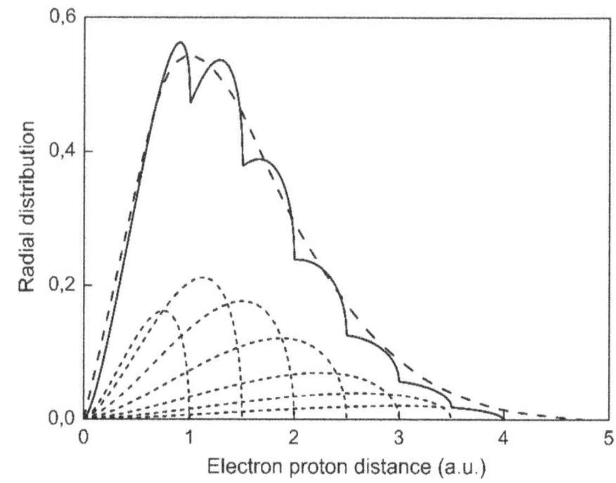

Fig. B.6 Radial distribution of an electron bound to a proton to form atomic hydrogen. Short dashed curves, individual radial distributions weighted by 8 appropriate coefficients a_i (see Table B.1); full curve, sum of the 8 distributions; dashed curve, quantum radial distribution

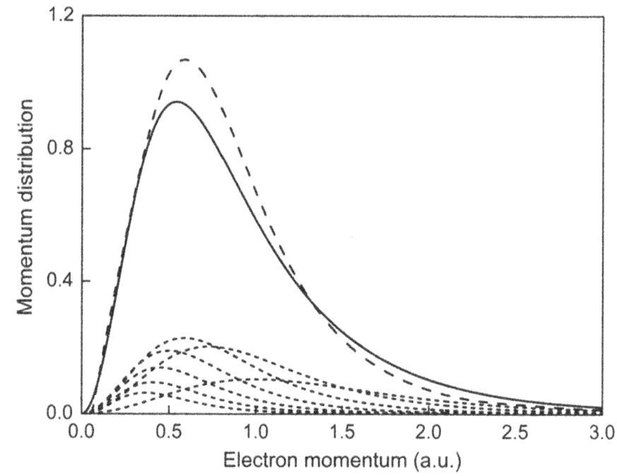

Fig. B.7 Momentum distribution of an electron bound to a proton to form atomic hydrogen. Short dashed curves, individual radial distributions weighted by 8 appropriate coefficients a_i (see Table B.1); full curve, sum of the 8 distributions; dashed curve, quantum momentum distribution

with the condition of normalization $\sum_i a_i = 1$

In a similar way, the momentum distributions writes

$$\rho_E(p_e) = \sum_i a_i \rho_E(p_e, E_i) \qquad \text{(B.14)}$$

Using an adequate set of a_i values, the radial and momentum distributions are very close to the quantum ones. However, the binding energy is no more unique, but is constituted of several discrete values of energies E_i. An example of radial and momentum distribution is shown in Figs. B.6 and B.7 for eight individual distributions, and Fig. B.8 gives the corresponding energy distribution. The coefficients a_i are shown in Table B.1.

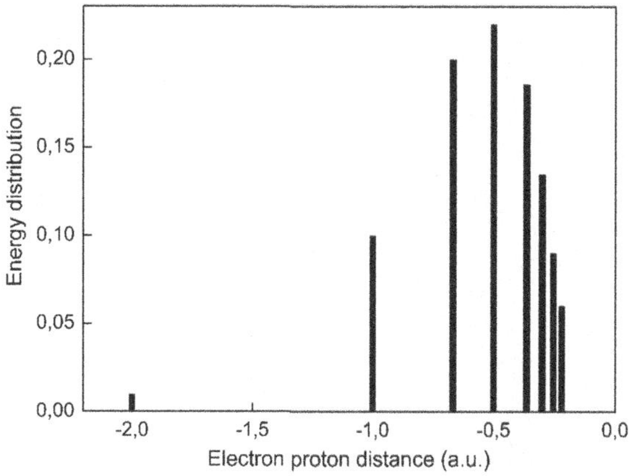

Fig. B.8 Energy distribution of an electron bound to a proton to form atomic hydrogen. The sum of the coefficients is equal to unity

Table B.1 Coefficients a_i used to weight individual radial distributions

E_i(a.u.)	a_i
– 2	0.01
– 1	0.1
– 0.667	0.2
– 0.5	0.22
– 0.365	0.186
– 0.333	0.135
– 0.255	0.09
– 0.22	0.06

B.3 Quasi-Continuous Radial Distributions with Discrete Charges

Very recently, a CTMC methodology has been introduced, especially conceived to provide a more accurate representation of charge-exchange processes between highly charged ions and hydrogen in its fundamental and excited states 2s and 2p [5]. This model is based on a more precise description of the target system than the microcanonical ensemble.

In contrast with the preceding model that introduced discrete values of electron energies, the present one considers different discrete values Z_i of target nuclear charges to reproduce the exact momentum distribution and the correct exponential tails of the radial distributions. By keeping the ionization potential as a fixed parameter, the quantum–mechanical radial distribution is approximated as a sum of classical radial distributions

$$\rho_E(r_e) = \sum_i a_i \rho_E(r_e, Z_i) \tag{B.15}$$

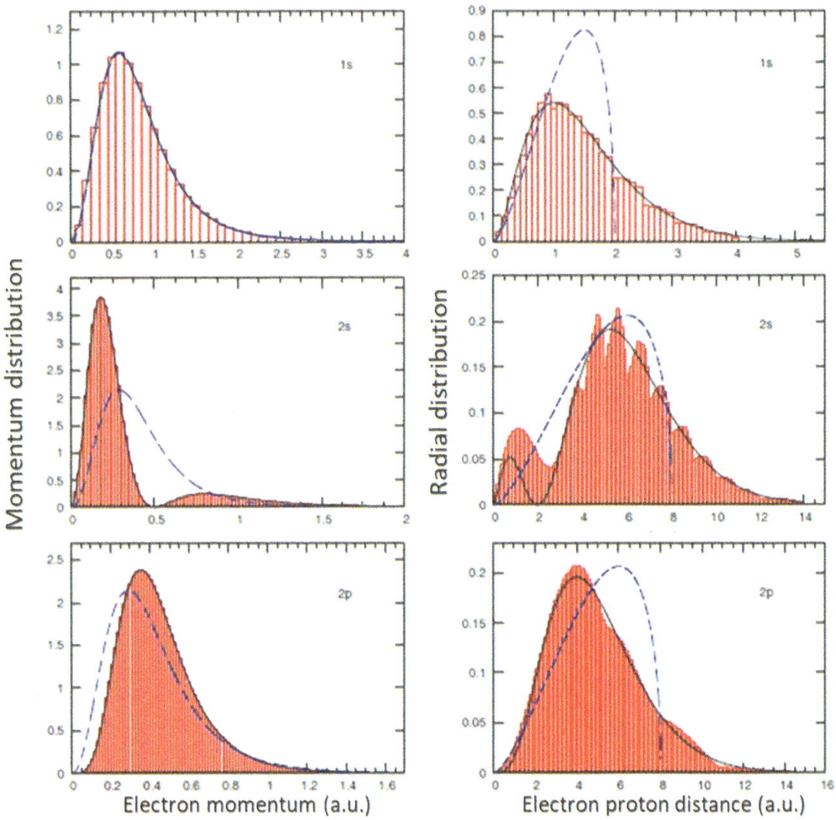

Fig. B.9 Momentum (left side) and radial (right side) distributions of an electron bound on a proton on its fundamental state (top of the figure), and on 2s and 2p excited states (center and bottom of the figure, respectively). Red bars, present model; full curves, quantum distributions; dashed curves, microcanonical distributions

with, again, the condition of normalization $\sum_i a_i = 1$. The values of a_i depend onto the initial orbitals considered.

Figure B.9 shows the momentum and radial distributions (red bars) derived from the present model, compared to the quantum distributions (full curves) and the micro-canonical distributions (dashed curves). It has to be noted that the radial distributions are remarkably reproduced, and that the nodal structure is qualitatively reproduced in the case of the 2s orbital.

References

1. Wigner, E.: On the quantum correction for thermodynamic equilibrium. Phys. Rev. **40**, 739 (1932)
2. Eichenauer, D., Grün, N., Scheid, W.: Classical trajectory calculations for H^+–H collisions with the Wigner function as initial phase space distribution. J. Phys. B **14**, 3929 (1981)
3. Sattin, F.: The electron capture cross section using the r-CTMC method: a review and a suggestion for improvements. In: Arnoldus, H.F., George T. (eds.) New Topics in Theor. Phys., Nova Science Publishers, p. 31 (2007)
4. Hardie, D.J.W., Olson, R.E.: Charge transfer and ionization processes involving multiply charged ions in collision with atomic hydrogen. J. Phys. B **16**, 1983 (1983)
5. Cariatore, N.D., Otranto, S., Olson, R.E.: Classical description of H(1s) and H^* ($n = 2$) for cross-section calculations relevant to charge-exchange diagnostics. Phys. Rev. A **91**, 042709 (2015)

Appendix C
Evolution of Integration Precision with Variable Time Step

To determine particle trajectories, Runge–Kutta method is used with a variable time step (cf. Chap. 2). The time step is a function of the different potentials acting on each particle and is proportional to a factor ε that is fixed, in most of the problems considered, to 0.5. From two examples, we will see that the value of ε is crucial in the precise determination of the final quantities, such as projectile and target kinetic energies.

C.1 Two-Body Problem: C^{6+} + He^{2+} Collision

Suppose a fully stripped C^{6+} colliding with a fully stripped target He^{2+}, for a projectile energy of 10 keV, for an impact parameter $b = 0.1$ a.u. During the collision, the only process that occurs is elastic scattering. Suppose both particles in a plan (xOy). Figure C.1 shows both particle trajectories for two values of ε ($\varepsilon = 0.5$, full curve; $\varepsilon = 0.9$, dashed curve).

It is clearly seen that, for the largest value of ε, the trajectory strongly deviates from the other trajectory at the earliest stage of the collision. For values of ε smaller than 0.5 (not shown in Fig. C.1), the trajectories become mixed up with that obtained using $\varepsilon = 0.5$.

The limit value of 0.5 for ε is confirmed when looking for the final energies of the projectile (circles in Fig. C.2) and of the target (squares in Fig. C.2). For values of ε smaller than 0.5, final energies are constant (7758 eV and 2239 eV for projectile and target, respectively). At higher ε values, energies noticeably deviate from the expected values.

A more reliable quantity to represent is the relative uncertainty $\Delta E/E_{exp}$, with $\Delta E = \left|E_{calc} - E_{exp}\right|$, where E_{calc} and E_{exp} are the final calculated and expected energies of the projectile, respectively (Fig. C.3). For values of ε smaller than 0.1, $\Delta E/E_{exp}$ is constant and of the order of a few 0.01%. Between 0.1 and 0.5, $\Delta E/E_{exp}$

© The Editor(s) (if applicable) and The Author(s), under exclusive license to Springer Nature Switzerland AG 2021

F. Frémont, *Classical Treatment of Collisions Between Ions and Atoms or Molecules*, Springer Series on Atomic, Optical, and Plasma Physics 118,
https://doi.org/10.1007/978-3-030-89428-3

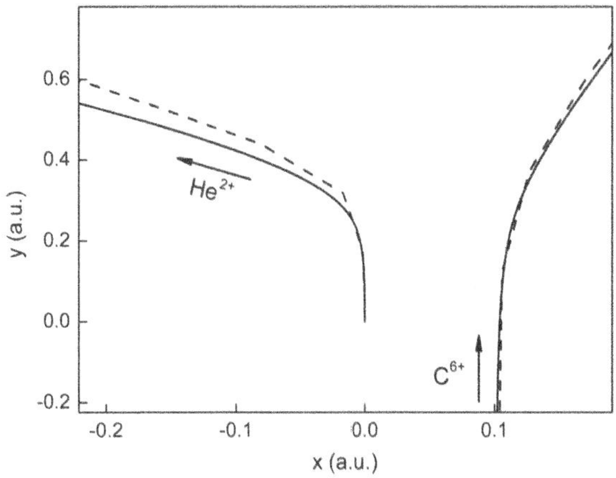

Fig. C.1 Trajectories of C^{6+} impacting on He^{2+} target, calculated using Runge–Kutta method and a variable time step ($\varepsilon = 0.5$, full curve; $\varepsilon = 0.9$, dashed curve)

Fig. C.2 Final projectile and target energies (full circles and full squares, respectively) calculated using Runge–Kutta method of order 4, as a function of parameter ε. The dashed curves represent the expected final energies

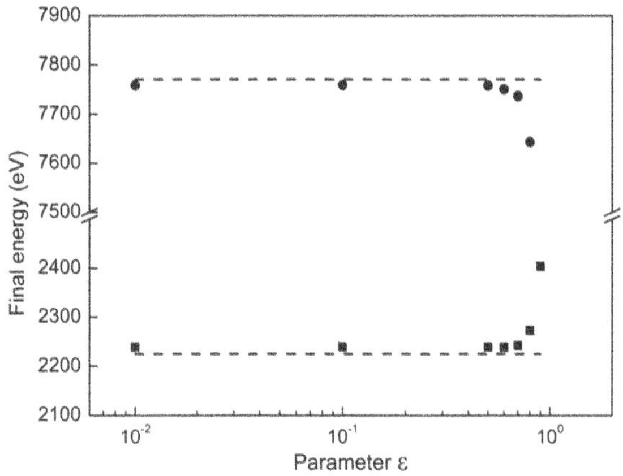

Fig. C.3 Relative variation of energy $\left|E_{calc} - E_{exp}\right| / E_{exp}$, where E_{calc} and E_{exp} are the calculated and expected projectile energies, respectively, as a function of the parameter ε

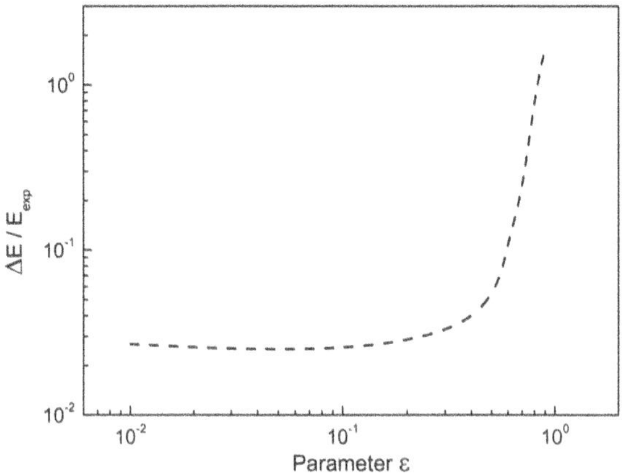

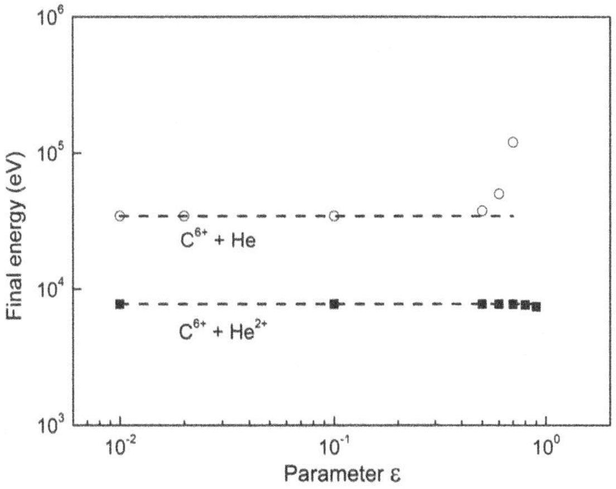

Fig. C.4 Final energies of a projectile C^{6+} following 10 keV $C^{6+} + He^{2+}$ (full squares) and 10 keV $C^{6+} + He$ (empty circles) collisions, as a function of the parameter ε. The dashed lines are the expected values of the energies. The energy in the case of $C^{6+} + He$ collision has been multiplied by a constant factor to separate both results

slowly increases. At larger values of ε, the relative uncertainty rapidly increases. Consequently, one would have to choose $\varepsilon = 0.1$ for this type of calculations. However, since only one process occurs, $\varepsilon = 0.5$ is sufficient to treat the collision.[1]

C.2 Three-Body Problem: C^{6+} + He Collision

The collision system $C^{6+} + He$ may be seen as a three-body problem if we consider that the two He electrons are independent from each other. Suppose, as in the preceding case, an elastic scattering, for a projectile energy of 10 keV. On Fig. C.4 are plotted the calculated final projectile energies of C^{6+} in the case of He^{2+} (full squares) and He (open circles) targets, as a function of ε. The final energy for He target has been multiplied by a constant factor in order to separate the results. At first glance, it appears that below 0.5, the final energy remains constant, as for the two-body problem. Consequently, a value of 0.5 for ε seems to be reasonable.

However, if we get into the details, we see strong differences between two-body and three-body collision calculations. More precisely, the quantity $\Delta E/E_{exp}$, represented in Fig. C.5 (full curve), clearly shows that value of 0.5 used for two-body collisions is no more efficient for three-body problems, since the uncertainty is of the order of 30%. Consequently, one has to take smaller values of ε (typically 0.1 at maximum) to decrease uncertainties in the total, partial and differential cross sections.

---

[1] Note that the sum of both energies is not equal to 1 keV. The small difference is due to the fact that, in the calculation, the projectile begins at close distances from the target (20 a.u.).

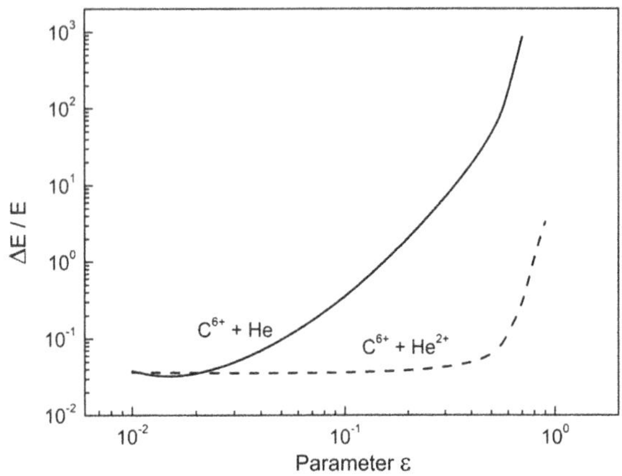

Fig. C.5 Relative variation of energy $\left|E_{calc} - E_{exp}\right|/E_{exp}$ for 10 keV $C^{6+} + He^{2+}$ (dashed curve) and 10 keV $C^{6+} + He$ (full curve) collisions, as a function of the parameter ε

Index

© The Editor(s) (if applicable) and The Author(s), under exclusive license to Springer Nature Switzerland AG 2021
F. Frémont, *Classical Treatment of Collisions Between Ions and Atoms or Molecules*, Springer Series on Atomic, Optical, and Plasma Physics 118,
https://doi.org/10.1007/978-3-030-89428-3

T

U

V

W

X

Z

The manufacturer's authorised representative in the EU is Springer Nature Customer Service Centre GmbH, Europaplatz 3, 69115 Heidelberg, Germany. If you have any concerns regarding our products, please contact ProductSafety@springernature.com

Printed and bound by CPI Group (UK) Ltd, Croydon, CR0 4YY
30/09/2026
02225254-0011